AF879942

Shorts Aircraft of WWI

Volume 1 | Early Types

Colin Owers

Great War Aviation Centennial Series #61

This book is dedicated to Gordon Bruce who made available to me his research work on Shorts aeroplanes. Gordon has one of the most organized minds that I have encountered and his ability to view the big picture has resulted in a much better book than I would have presented on my own.

Acknowledgements

The late J.M. Bruce wrote that a lot depends on whose reference books you read. Thanks to friends old and new, some sadly no longer with us, the scope of research for this book has encompassed much more than I could ever read myself. Thanks are due to. Gordon Bruce; the late J.M. Jack Bruce; Moshe Bulhman; Ian M Burns; Andrea Casarrubea; Peter Cowlan; AVM Peter Dye, RAF (Ret); Mark Evans; Roberto Gentilli; Frits Gerdessen; Knut Eric Hagen; Jack Herris; the late Barbara Huston; Colin Huston; Sir Peter Jackson, P. Phil Jarrett; Emin Kurt; Mark Lax; the late G. Stuart Leslie; Peter London; Geoffrey Miller, author of *Superior Force - The Conspiracy behind the Escape of the Goeben and Breslau*; Ivan Siminic; W. Bill Toohey; R. Ray Vann; Paolo Varriale; the Aviation Heritage Trust (AHT); the staff of the National Archives of the UK and USA; the RAF Museum, Hendon; the FAA Museum, Yeovil; the USN Naval History and Heritage Command (NHHC); the US National Naval Aviation Museum, Pensacola; the NASM, Smithsonian Institute; the RAAF Museum, Point Cook. Special thanks to my wife Julie who copied many of the documents used in developing this book. Any errors are mine alone. Corrected version.

Color aircraft profiles © Bob Pearson. Purchase his CD of WWI aircraft profiles for $50 US/Canadian, 40 €, or £30, airmail postage included, via Paypal to Bob at: **bpearson@kaien.net**

For our aviation books, please see our website at: www.aeronautbooks.com.

I am looking for photographs of the less well-known German aircraft of WWI to complete this series. For questions or to help with photographs you may contact me at jherris@me.com.

ISBN: 978-1-953201-44-7

Design and layout: Jack Herris
Cover design: Aaron Weaver
Digital photo editing: Jack Herris

Aeronaut Books

Books for Enthusiasts by Enthusiasts

www.aeronautbooks.com

Table of Contents

Volume 1

Volume 2

Volume 3

Introduction

The designation of the various early Short types varies amongst references. That adopted by R Sturtivant and G Page in their magnificent *Royal Navy Aircraft Serials and Units* (Air Britain, UK, 1992) has been followed in this history. The Royal Flying Corps (RFC) was divided into a Military Wing and a Naval Wing. The Royal Naval Air Service (RNAS) did not come into being until 1 July 1914, and lasted until the amalgamation of the RNAS and RFC to from the Royal Air Force (RAF) on 1 April 1918. The term RNAS is used throughout until the units became RAF units.

Shorts assigned an 'S' number to their early orders and these also covered spares, repairs, etc. From the time that Shorts began to construct Short-Sommer type boxkites, each Short aircraft received a construction number (c/n) with a 'S' prefix, further complicating things.[1] The later 'S' construction numbers are not always dependable as several early machines were extensively rebuilt. Thus the S.56 was an improved S.41.

Frank McClean gave his machines his own number and some of these were subsequently given an Admiralty serial number. There were three Admiralty serial systems, the first introduced in 1911 and changed twice in 1912. These Admiralty numbers were superseded on the formation of the RFC on 13 May 1912, with a Military and a Naval Wing.[2] The vague reference to what particular Short aircraft in the press and official records in the early days, as well as during the war, makes difficult the task of determining which aircraft was being discussed. Another feature of the early RNAS types was their retention of their original serial even when they were 'repaired' or rebuilt and returned to the air looking nothing at all like they did when they entered the workshop. The British Official History, *The War in the Air,* refers to Short seaplanes without reference to their type but sometimes does refer to their horsepower. It was common for the machines to be referred to by their horsepower in official documents. Thus, the official *Truing-up of Aeroplanes 1916* published by the Air Department of the Admiralty in 1916, showed the diagrams for the *Short 150 H.P. Seaplane; 135 H.P. Canton-Unne Seaplane Short; 200 H.P. Canton-Unne Seaplane Short* and *Short 225 H.P. Sunbeam sea plane.* There is no other mention of any designation as to the Type. The drawings for the 135-hp and 200-hp Canton Unné seaplanes are the same with the data changed.[3]

In addition, the Admiralty had a system of aircraft nomenclature that is hard to understand. Sometimes the type was named after the serial of the first aircraft of the production run, such as the Short 74. Frequently the number seems to have been chosen at random or the serial of the aircraft that carried out the performance testing for the type, such as the Short Types 166 and 827. Then the Admiralty made a unique departure in naming the Short 320 – naming it after the horse-power of its Sunbeam engine. This aircraft is also referred to as the Short 310! To make the matter of determining what type of aircraft is being discussed in official documents even more complicated, the official designations were not used, but the machines referred to by the horse-power of their engines! As J.M. Bruce noted: *The effectiveness at the times of its use of this somewhat tortuous method of providing aircraft designations is not now known, but it had undoubtedly caused a great deal of confusion, uncertainty and inaccuracy in latter-day references to the aircraft concerned.*[4]

In 1916 the Admiralty issued a Nomenclature of Types in order to *avoid misunderstandings or the alternative lengthy description* of a machine.

RNAS Nomenclature of Types

RNAS General Memorandum No.53 of 14 June 1916.

Aeroplanes

Breguet de Ch	Breguet de Chasse
Breguet de B	Breguet de Bombardement
Breguet Conc	Breguet Concours
Short Bomber	Short Land Machine
Sopwith 9901 type	Single seat fighting biplane.
Sopwith 9400 S type	1½ Strut. fighting type, short distance.
Sopwith 9400 L type	1½ Strut. fighting type, long distance.
Sopwith 9700 type	1½ Strut. bombing.
Sopwith 806 type 1	150 Sunbeam type.
Sopwith Scout	80 Gnome (obsolete type).

Seaplanes

Short 74 type	100 Gnome (obsolete type).
Short 184 type	225 Sunbeam.
Short 184 single type	225 Sunbeam, fitted with bombs in passenger's seat.
Short 827 type	150 Sunbeam
Short 830 type	135–140 C'Unné.
Short 166 type	200 C'Unné.

Source: TNA AIR1/2097/207/15/2.

Endnotes 1. Gordon Bruce noted that the aircraft were identified in the Shorts' Order Book job numbers, such as Colman's aircraft that was initially identified as Job No. 18 and after it had flown, as S.26 *under new regime.* To further complicate matters, the c/n started at S.26. Letter from G Bruce to Author, 19.09.2020.

2. See Sturtivant & Page for a reconciliation of Shorts Construction Numbers with Naval Serials. *Op Cit.* P.460.

3. This manual was reprinted by the Naval and Military Press Ltd.

4. Bruce, J.M. 'The Short Seaplanes,' *Flight*, 14 December 1956, P.924. Jack Bruce was unfortunate in that when he wrote this history the Shorts' Order Book had not been recovered. It was mislaid sometime after March 1943 and did not re-appear until after Oswald's death in 1969, and then was in private hands. It was reacquired by Shorts in the mid-1970s and is now in the Ulster Transport Museum, Cultra, Northern Ireland.

Chapter 1: The Short Brothers

Above: This photograph was known to Shorts as The Founding Fathers photograph. It marked the visit of Wilbur and Orville Wright to the Aero Club's flying ground and to the Short Brothers' adjacent factory at Eastchurch. The group is situated in front of Mussel Manor, the Club House. From left to right; Seated: Mr J.T.C. Moore-Brabazon; Mr Wilbur Wright; Mr Orville Wright, and the Hon C.S. Rolls. Standing: Mr J.D.F. Andrews, Mr. Oswald Short; Mr Horace Short; Mr Eustace Short; Mr Frank McClean; Mr Griffith Brewer; Mr Frank Butler, Dr W.J.S. Lockyer; Mr Warwick Wright. (via G Bruce.)

The name of Short Brothers will be forever associated with flying boats including the magnificent Sunderland of WWII fame. In WWI however, Short Brothers produced the majority of floatplanes used by the RNAS. 63 of the first 200 Naval aircraft that were ordered, or allotted serial numbers by the Admiralty, were Short designs.[1] However, the connection between the Short brothers and aviation goes back to 1901 when they started to manufacture balloons.[2]

There were three Short brothers, Horace Leonard (1872–1917), Albert Eustace known as Eustace (1875–1932), and Hugh Oswald known as Oswald (1883–1969).

Horace was struck with meningitis following a head injury while an infant. This resulted in an abnormal brain development that gave him a large head, but compensated by making him a genius.[3] He had an extraordinary early life, travelling to Australia by working his way on a sailing ship, continued through to South America and ended up in Mexico in charge of a silver mine where Eustace found him. Eustace had been sent by their mother to bring him home as their father had died in 1891 leaving the family in dire circumstances.

Unwilling to leave the mine, Horace gave his younger brother £500 and stated he would be home later. In 1896 Horace returned to England as he had invented a device that

Above: Workmen in Shorts' original aeroplane factory.

used steam or compressed air to amplify sound. He patented the Auxetophone in 1898.

Air Commodore E.L. Gerrard recalled that Horace Short taught the first four Naval aviators the theory of flight, etc. *Horace was very serious over anything to do with flying but gave full rein to humour betweenwise. His favourite quip was the invention of ridiculous words some of which passed into the language and are now in the dictionary e.g. Blimp for a non-rigid Airship. He pretended the word was invented by his small son.*[4]

Eustace and Oswald began to manufacture spherical hot air balloons for sale in 1901. Horace had flown with Eustace in a second-hand Spencer balloon and considered it unsafe and declined to join the brothers in this venture. He did allow them to use the loft above his premises where they built their first spherical balloon. In 1905 Eustace and Oswald were contracted to build three balloons for the British Army. So impressed with their work was the Royal Balloon Factory's superintendent, Col Templer, that he introduced the brothers to other balloonists, such as Charles Rolls, who then ordered a balloon to use in the 1906 Gordon Bennett balloon race. When Horace moved to Newcastle the other brothers were without a workshop until they located two arches under the LBSC Railway at Battersea Park, conveniently located near the local gas works. This move was helped by C.S. Rolls (of future Rolls-Royce fame). The pair came to the notice of the Aero Club, again with the support of Rolls, and they proceeded to build the majority of balloons for members.

In the 1908 Gordon-Bennett race Lt Barrington Kenneth and Eustace Short left Battersea at 10 o'clock on the night of 10 September in one of the Aero Club's balloons. The wind blowing at 30 mph from the north-west, they had every reason to believe that they would cross the Channel and arrive in Belgium via Ostend in four hours.

At 8 a.m. they were at 12,000 feet sailing into a towering bank of clouds that caused them to sacrifice two more bags of ballast leaving only seven from the 22 bags that they had started with. They flew into another bank of clouds and were caught in a snow-storm. The balloon immediately was covered in snow and condensation causing the balloon to rapidly lose height. Two bags of ballast were dropped but did not stop the descent. They had checked to ensure that the balloon basket and its fittings were water tight and had provided themselves with life jackets.[5]

With three bumps the balloon landed in the North Sea. Ballast was thrown out and the balloon was soon riding 30 feet above the water with the sea anchor bringing their speed down to about six mph. Pilot boats soon arrived and No. 5 lowered its lifeboat as the sea was still rough. The aeronauts instructed the crew how to pick up the trail rope and the balloon was then towed to the Pilot steamer. The valve was opened and the envelope was allowed to fall into the sea resting on the basket. Eustace pulled the rip-cord and then followed Barrington Kennett into the lifeboat. After struggling to get the balloon on board due to the

nterior of Short Brothers' principal erecting shop, where the British-built Wright flyers are in course of construction. Although hardly established three months in their new premises, Messrs. Short are already employing 80 men. The above shop is 140 ft. long by 45 ft. wide.

Above: The Rochester seaplane works remained unchanged until 1936.

inexperience of the Pilot boats crew, the pair returned to Britain.[6]

On the Wright's demonstration of their aircraft in France in 1908, the brothers decided that ballooning was finished and that they would have to change to constructing aeroplanes. They needed Horace's genius and were able to persuade him to give up his employment and in November that year the partnership was registered at Battersea as Short Brothers with a capital of £600. They obtained the rights to build six copies of the Wright designs. Wilbur Wright did not have any plans and so Horace and Eustace went to Pau, France, to measure and make detailed drawings of the Wright machine. A tract of land was purchased on the Isle of Sheppey in Kent where they erected a factory to manufacture the six Wright machines that had been ordered by members of the Aero Club. Frank McClean had purchased the Eastchurch farm and leased it to the Aero Club. He kept a small cottage for himself as a week-ender. The Shorts factory was established on adjacent land and various member of the Aero Club had sheds on the airfield. Alec Ogilvie and McClean had one as did the Blair Atholl Syndicate. The Admiralty's property was situated at the west end. The whole establishment was to become the Eastchurch Air Station.[7] This established Short Brothers as the first aircraft production factory in the world. Charles Cyril Turner wrote that Horace Short was *the inspiring genius. Not only his development of the seaplane, and other types, but the sense of responsibility he had towards pilots, and the exacting care he took in aeroplane construction, helped to ensure the freedom of Eastchurch from fatal accidents in those days. He certainly laid the foundations of the British aviation industry.*[8]

Charles Rumney Samson, who will appear many times in this history, was also full of praise for Horace. *The name of the late Horace Short stands like a lighthouse in the middle of a shoal-infested sea; without doubt Aviation, not simply Naval Aviation, owes him a great debt. He was a man of vast intellect, determination and above all a sturdy support in times of doubt.*[9]

Writing in 1930, Oswald noted that in the pre-war days *Seaplane builders were fully "approved firms" in the real sense of the term, and were subject to no restrictions of any kind. It was wisely realised that it was necessary to learn to walk before trying to run, and the requirements, meagre as they were, were quite difficult to carry out. A top speed of 69 miles an hour was considered rather wonderful and a ceiling of 5,000 ft. wonderfully good.*[10] The Admiralty would purchase one or two machines, sometimes the two would be variations of the same design, in order to find, by trial and error, the best way to achieve a practical seaplane. To this end, officers of the RNAS would work with Short Brothers and other designers. When Horace died in 1917, Oswald, the youngest brother,

Above: Horace Short (1872–1917). The Royal Aeronautical Society described Horace as having a 'Boy's Own' early life before going on to build some of the first British Naval aircraft.

Above: Eustace Short (1875 –1932). It was Eustace's love of ballooning that led to the Short brothers becoming involved in aviation.

took over the company as Eustace was engaged in building airships for the Navy. Oswald soon proved himself well capable of his new responsibilities, his genius immediately asserting itself.

It was the love of Eustace, the middle brother, for ballooning that had caused the brothers to go down their path as constructor of airships and aeroplanes. While there was a demand for lighter-than-air craft, Eustace looked after this side of the business, that is up to the end of the war. It was Eustace who designed the mooring buoy adopted as standard by the RAF for mooring out its seaplanes. The buoy was a rubber ball, sealed except for a valve at its bottom. The

Right: Hugh Oswald Short (1883–1969). Known as Oswald, 'The Kid', as Horace referred to him, was the youngest of the Short brothers. C.C. Turner recorded that Horace was the authority on the flying machine generally, while his brother Oswald specialised on seaplane floats.

Nile River crocodiles found them tasty and the cure was found to be colouring them bright red. Eustace decided to learn to fly in 1925, he had alighted perfectly in the light seaplane Mussel II on the Medway on 8 April 1932, but the aircraft grounded in the mud on the Coxton side of the river. Eustace had died at the controls.[11]

Oswald continued to run Shorts until the company was acquired by the Government in 1943. He died in 1967. John Lankester Parker, Shorts chief test pilot for 27 years and a Company director, wrote that *The contribution made by Oswald to the technical side of aviation is, I think, far greater than is generally realised and in my view he has never received the rightful tribute for his imaginative and practical work.*

Oswald, personally, sponsored the use of aluminium alloy for the primary structure of aircraft years before the material was accepted as 'airworthy'.

He pioneered the stress-skinned, all-metal construction with the most fanatical enthusiasm.[12]

Short-Wright No. 6, the last of the type, was built in March1909 for C.S. Rolls and then obtained by Alec Ogilvie. The last two Short built Wrights had an improved four-boom tail outrigger with a single fixed tailplane mounted behind the rudder. Ogilvie modified his machine to the later Wright standard with improved control and fitted wheels alongside the skids. Further modifications saw a nacelle for the pilot and passengers. Remarkably this machine survived until 1915 when it was purchased by the Admiralty (Contract CP66325/15) from the then Sqn Cdr Ogilvy and given the RNAS serial No. 1373. Delivered to the Eastchurch school in August 1915 it was reported as damaged 2 October 1915, and deleted on 13 January 1916. Surely a record for such an early aircraft.

While the Wrights flew well, the first Short design, the McClean Short No. 1, was unsuccessful.[13] Frank McClean was responsible for many of the early Short designs and was also instrumental in establishing the RNAS. John Lankester Parker, the Shorts test pilot, recorded that *Frank McClean was a chronic customer.*[14] His contribution to the initial success of Short Brothers cannot be underestimated.[15] Short No. 2 was based more on Wright lines and won the *Daily Mail's* prize for the first aircraft to make a flight of one mile in England by a British pilot in an all-British machine.

In December 1910, the weekly journal *Flight* announced that *the Committee of the Royal Aero Club has placed two aeroplanes at the disposal of the Admiralty for the use of Naval Officers at Sheerness and Chatham, at the Club's flying grounds at Eastchurch, Isle of Sheppey. The machines are biplanes, fitted with Gnome motors. Several members have promised to assist in giving practical instruction to the officers.*[16] Admiral Sir Charles Drury, the Commander-in-Chief at Nore, issued a General Fleet Order in November 1910, signalling the Admiralty's intention to accept the offer.

At this time, thanks to the great generosity of Frank McClean, the first four Naval Officers, Samson, Gregory, Longmore and Gerrand, learned to fly at Eastchurch.[17] The four volunteers were chosen out of over several hundred. Grace had been lost in the Channel and his place as trainer was taken by George Cockburn.[18] In addition to their flight instruction Horace Short gave each six months technical training for a cost of £20 and all breakages, although the Admiralty stated that it 'might' see fit to pay the fee for the students. The selected four, three naval officers and a marine, were: Lts R Gregory[19], Charles Rumney Samson[20] and Arthur Longmore,[21] all Royal Navy, and Lt Gilbert Vernon Wildman-Lushington, RM, Artillery. Wildman-Lushington was on sick leave when instruction commenced and his place was taken by Lt Eugene Louis Gerrard, RM, LI. All qualified in April and May1911. The initial offer was for the use of two of McClean's aeroplanes, however, through a close association with Short Brothers, a number of different models were trialled. The four volunteers reported to Eastchurch on 1 March 1911.

McClean went on a solar eclipse expedition to the Tonga Islands in the first half of 1911. He left instructions with Horace Short that his Short machines S.26 and S.28 plus his new Short (S.34) should be used to train the Naval applicants. In addition, McClean provided the cost of employing a competent mechanic to look after the machines as well as to provide for any 'reconstruction' that may have been necessary to these second-hand aeroplanes. Thus, when S.26 became unsafe to fly, it was replaced by the S.34. When S.28 required reconstruction, the work was so extensive that Shorts gave it a new c/n and it emerged as the S.38. The Navy aviators also had access to the S.32 that had passed to McClean after Grace's disappearance, as well as McClean's other aircraft and Maurice Egerton's S.35.

Fortunately, Gregory has left an account of his training.

The Naval Aviation Course commenced on March 1st, 1911 at Royal Aero Club Flying Ground at Eastchurch.

Mr Cockburn came down to instruct officers on March 14th and no flights took place prior to that date.

Between March 15th and about April 10th did 10 passenger flights with Mr Cockburn, in order to get accustomed to handling the controls. The duration (total) of all passenger flights was 1½ hours, & distance covered about 60 miles. The passenger flights were very much hampered by bad weather, there being intervals of as much as 10 days, on which flying was impossible.

Made first flight (straight) by myself on April 22nd. A first circular flight on April 26th and qualified as pilot under 1911 Rules at IV.15 a.m. on Monday 1st May 1911. On May 2nd at General Meeting of the Royal Aero Club Committee was awarded "Aviators Certificate No. 75" under the regulations of Féderation Aeronautiqué Internationale.

Gregory had a photograph of the Farman type of machine pasted into his Log Book with the caption *Machine on which certificate was flown for. Short Bros (Farman Type) Biplane No. 34 fitted with 50 h.p. "Gnome Engine."*

The quartet attended the French Military Aviation Trials at Reims and various French factories, including the Gnöme engine factory, as they were considered experts in aeronautics at the end of their course.[22]

Samson recalled that after the six months' course had been completed the four newly minted aviators *were left like fish out of water.* The Admiralty had no ideas of what to do with them. Samson states that he summoned the resolve to meet with Adm A.K. Wilson, the First Sea Lord. He asked that the Admiralty buy the two aeroplanes that they had learnt to fly on and let them go on flying. He required four Engine Room Artificers, four Shipwrights and four Bluejackets. Also, he wanted permission to order a hydro-aeroplane from Shorts *to see if it was possible to fly off the sea.* He promised to prove it was possible to fly off a ship within six months. Surprisingly, considering Adm Wilson's aversion to aviation, all his requests were granted.[23]

In his post-war recollections Samson states that *We went on flying McClean's Aeroplanes until October 1911, and then I went up to the Admiralty with proposals. They bought two aeroplanes from McClean and sent me 12 Naval ratings to start a School and experimental work. I was put on the Committee of Imperial Defence to discuss and propound schemes for Naval and Military Aviation.* In typical Samson fashion he wrote that *I was the only practical flying expert on this Committee, and a certain amount of credit should be due to me.* He was a member of the Air Committee and when the RFC was formed in May 1912, he was head of the Naval Wing.[24] The Navy obtained a permanent station at Eastchurch that would allow for training on landplanes and seaplanes with ground-staff being trained in the Short Brothers' workshops. From this point on the work of Shorts and the Navy became intertwined as the two worked together to produce machines that met Navy requirements.

The 17 October 1912, issue of *The Aeroplane* described 'Developments at Eastchurch', and includes an aerial photograph showing the Naval and Royal Aero Club sheds and the Short Brothers' factory that was reported as *increasing rapidly in size.* By 1914 Short Brothers needed to expand and new works were built at Rochester, just above the Rochester Bridge, on the banks of the Medway River.[25] *Flight* described the facility in detail in its issue of 7 March 1914.[26]

By 1912 the Short Brothers had produced a number of different aeroplanes – monoplanes and biplanes, pushers and tractors, some with wheels and others with floats. According to John Lankester Parker, they were great believers in seeking more power and made the first twin-engine aeroplane in the world.

The Standing Sub-Committee of the Committee for Imperial Defence presented a list of aeroplanes, hydro-aeroplanes and floats (safety device for flying over water) in February 1912. Samson was the Navy's representative on the Committee as noted above. This listed eight aeroplanes and 14 hydro-aeroplanes. The latter being two Short machines, one a biplane and one a monoplane with twin engines to be ordered at once, six to be tendered for by English firms, the further six to be ordered from the firm making the most satisfactory machine. The hydro-aeroplanes *were not*

Course of Instruction for Naval Officers in the Management and Piloting of Aeroplanes

On 16 February 1911, Horace Short wrote to the Navy enclosing a set of instructions for the proposed training of Naval officers. It would cover their manufacture and design, and include a series of lectures by Mr H.L. Short on the theoretical and practical application of aerodynamics to aerial flight.

Two biplanes were to be made available to the students and the students should practice on them alternatively so that the four shall have the same hours while the good weather lasted. A third machine was nearing completion and would be available in about two weeks time. A fourth machine is also under way, with duplex power plants which will also be at their disposal when completed.

Suggested Course of Instruction for Naval Officers.

Hours to be from 10 a.m. to 1 p.m. and 2 to 4 p.m. on the first five days of the week. However, if the weather is good, Saturday and Sunday may be used for actual flying.

For the first two days, the period up to lunchtime would be Grounds workshop experience. ...a general examination of complete machines and their separate parts.

After the students received six lectures they would be well grounded in the conditions which they will encounter in the preliminary flights, and will then be sufficiently acquainted with the construction of the machines to enable them to confidently undertake the first lessons in actual flight.

Flying will take place on all possible occasions, weather permitting. When flying was not possible the students would be in the workshops. They can assist in the disassembling of the Gnôme engines.

Before taking a machine out the student will examine all the principal Wires, Controls and fastenings.

In the case of any breakages to the machine which Students are using they will attend to and observe the method of such repairs as are necessary, and keep a log of the same with details.

The Student will also keep a log of his flying lessons, and of any flights he may make up to the sixth month; this log to contain such information as weather conditions, velocity and direction of the wind, atmospheric temperature and barometrical pressure. He would measure the quantity of lubricating oil and fuel in the tanks and take the maximum speed of the engine fully opened out whilst resting on the ground. He will also log the time of starting away to the time of his landing and will make a mental note whilst in flight of the course he is steering and of the various movements of the rudder which he may have to make due to wind currents.

The students were also to be made familiar with the Statescope, Barograph and Aneroid as used for aeronautical purposes.

Source: TNA ADM1/8212

included in the interim report, as their immediate supply was not considered to be a matter of such vital urgency as in the case of the aeroplanes. As experiments with hydro-aeroplanes are in their infancy and very few successful flights from water have been made at all, and these few only from still water, the Sub-Committee recommend that the purchase of the six hydro-aeroplanes referred to … should be conditional on successful experiments with the two Short machines.[27]

Around April 1912, the public was notified that the purchase of 12 aeroplanes, hydro-aeroplanes and floats for first requirements under this plan had been undertaken. *It has been deemed desirable to test a number of types with a view to arriving at the most suitable pattern for naval services. As soon as a satisfactory type of aeroplane for future use has been evolved, orders will be given so as to ensure that there are always not less than two machines of the same type available.*[28]

The success of Shorts was due to their willingness to work with Samson and the Navy. Making their facilities available for the training of the original Navy pilots was paying big dividends. The Admiralty wrote to Winston Churchill that Lt C.R. Samson recommended that the orders for *Experimental and Fleet work* should be given to Short Bros to construct immediately. *Short Bros. are the only makers of twin engine machines. These machines will be made at Eastchurch under our supervision. Shorts is the best work in England so success should be gained with both machines* – the twin-engine hydroaeroplane biplane and the twin-engine hydro-plane monoplane.[29]

In the event, the twin-engine hydroplane biplane was abandoned as it was too ambitious. The only aircraft in the Short Order Book that Gordon Bruce's research has been able to connect to the proposed 'at once' orders is S.46, built on Job No. 232, H2 – the 'Double Dirty.'[30]

By the end of 1912, the Navy had the following aircraft

# of	Type	RNAS Serial
2	50-hp Gnôme Short biplanes for School work.	T1/1 & T2/2.
2	Twin 50-hp Gnôme Short Biplanes	T3/3 & T4/4.
1	70-hp Gnôme Short tractor.	T5/5
1	50-hp Gnôme Short monoplane.	M2
1	100-hp Gnôme Short Seaplane	H1/10
1	Twin 70-hp Gnôme Short monoplane	H2/12
1	110-hp Canton-Unné Breguet biplane	T6/6
1	70-hp Gnôme Deperdussin monoplane	M1/7
1	50-hp Gnôme Nieuport monoplane.	M3/13
1	65-hp Austro-Daimler Etrich monoplane.	M4
1	70-hp Gnôme Henri Farman seaplane	11
1	80-hp Gnôme Donnet Lévèque seaplane	H7/18
1	50-hp Gnôme Bristol for school work.	24
1	70-hp Gnôme Sopwith biplane	27
1	Willows airship.	

Source: "Naval and Military Aeronautics – 1913." TNA AIR1/2126.

Endnotes

1. Manson, T. "Royal Naval Air Service Aeroplanes of the Royal Flying Corps (Naval Wing) and RNAS 1911 to 31 August 1915." *Cross & Cockade (Great Britain),* Vol. 15 No.4 1984. P.145.
2. Oswald Short's Diary gives the year as 1901, not the 1902 usually quoted. Research by G Bruce.
3. Horace suffered from *fietus in foetu* – two brains had developed in his head. This was discovered in the post mortem after his death from a brain haemorrhage. Driver, H. *The Birth of Military Aviation – Britain, 1903-1914,* The Royal Historical Society, UK, 1997. P.42.
4. E.L. Gerrard. "Early Days of Flying." FAA Museum 1997/207/022.
5. The 17 October 1908, issue of *The Airship* contained an article on the Short Brother factory under the railway line. Describing the basket – car the article noted that *the padding has been made to serve yet another purpose, for one of its qualities is that it gives buoyancy to the car, rendering it absolutely unsinkable in the event of a descent into the sea.* The cars were not watertight, meaning that they were built to allow water to flow out of the car, otherwise if it shipped water, then its buoyancy would disappear and it would be almost impossible to get rid of the water.
6. At this time the Channel had been crossed by balloon 32 times since 1785. There had been a few fatalities, the first being a sailor named Prince who tried to escape the Germans besieging Paris in November 1870, but came down near Land's End. This section has been written from "An Adventurous Balloon Voyage – Dropping into the North Sea." *The Airship,* November 1908. P.32. Also "Aeronauts Rescued in the North Sea" *The Times*, 12 September 1908. This may have been a bigger story except for the wedding of Winston Churchill in the same edition.
7. For the story of Eastchurch see Empson, R.H.W. "Eastchurch: The Early Years – A History of the Eastchurch Air Station, Sheppy 1909-1920." Reprinted

in *Cross & Cockade International,* Vol. 40 No. 2, 2009. P.71.

8. Turner, C.C. *The Old Flying Days,* Samson & Low, Marston & Co Ltd, UK. 1927. P24. Turner has been called the first aeronautical correspondent and was on hand reporting on most of the early activities taking place in the 'Old Flying Days.'
9. 'Early British Aviation.' This undated paper was written by the then Air Commodore Samson after his retirement. Copy in IWM via G Bruce. Richard Bell Davies served under Samson and wrote that anyone who served under him under six months never had a good word to say about him. Anyone who served over six months never had a bad word to say about him. *His outstanding quality was guts…His manners were brusque, he was frequently rude and had no tact, but as a friend he was absolutely loyal.* R Bell Davies, *Sailor in the Air,* Peter Davies, UK, 1967. P.75.
10. Short, O. 'Twenty-one Years' Progress of Seaplane Design,' *Flight,* 3 January 1930, P.67.
11. Death of Eustace Short,' *Flight*, 12 April 1932, P.332.
12. 'Short Borthers Aviation Pioneers – Hugh Oswald Short.' Shortbrothersaviation pioneers.co.uk.
13. Hugh Driver noted in his *The Birth of Military Aviation, Britain 1903-1914* (The Royal Historical Society, The Boydell Press, UK, 1997), mentions the fact that the machine built for Moore-Brabazon was a glider built to Moore-Brabazon's design, and it was he who installed an engine to the airframe. The machine did not fly, leading A.V. Roe to claim in his 1939 autobiography that the Short-built machine had been ineptly designed. Oswald Short denied that he ever knew that Moore-Brabazon planned to fit an engine to the airframe. The controversy led to the three, Roe, Short and Moore-Brabazon, conducting *an acrimonious exchange of letters* in *Flight* and *The Aeroplane* in 1940. Driver, H. P.47.
14. Lankester Parker, J. 'The First Short Brothers Commemorative Lecture'. Reprint from *the Journal of the RAeS*, Vol. 66, No. 63, January 1962.
15. See Jarrett, P, *Frank McClean: The Godfather of British Naval Aviation,* (Seaforth Publishing, UK, 2011), for a detailed biography of McClean. One of the original four Naval aviators, E.L. Gerrard, wrote that the Admiralty was very disinclined to start the difficult process of getting the Navy into the air, *but their hand was forced by Frank McClean.* He wrote that McClean was a man *with a conscientious objection to a Bank Balance, he thought it to the general benefit that money be kept circulating: aviation helped a lot in putting his theory int* (sic) *practice.* 'Early Days of Flying.' FAA Museum 1997/207/022.
16. 'The Club's Offer to the Admiralty' in the 'Royal Aero Club of the U.K.' column. *Flight,* 3 December 1910. P. 992.
17. Lankester Parker, J. 'The First Short Brothers Commemorative Lecture'. Reprint from *the Journal of the RAeS*, Vol. 66, No. 63, January 1962.
18. Cockburn himself had described Grace as *probably the best pilot of those days.* Quoted in Driver, H. *Op Cit.* P.71.
19. Gregory was in command of Great Yarmouth station on 4 August 1914. He had seven officers, two warrant officers and 40 men. Aircraft initially comprised one Maurice Farman and one Short aeroplanes, one Henri Farman and one Short plus four Sopwith seaplanes for a total of eight machines. Hobbs, D. *Royal Navy's Air Service in the Great War,* Seaforth, UK. 2017. P.66.
20. Samson published his rollicking memoirs as *Flights and Flights,* (Ernest Benn, London, 1930). Unfortunately, he does not touch on his pre-war experiences. E.L. Gerrard recalled that Samson crashed his aircraft when taking off for the Mortimer Singer competition for service personnel. *Samson fitted enormous extensions to the wings of his m/c and crashed on take off; I never heard such a crash! The splintering of wood seemed to last for minutes (Samson could always be relied on to convert everything he did into the spectacular.)* 'Early Flying Days.' *Op Cit.*
21. Longmore (later ACM Sir A.M. Longmore, GCB, DSO), published his autobiography as *From Sea to Sky 1910-1945,* (Geoffrey Bles, London, 1946). E.L. Gerrard wrote that Longmore stated that he had been selected *because he was regarded as expendable, and would leave no widow to claim a pension! I think his good looks and tactful bonhomie must have helped.* 'Early Flying Days.' *Op Cit.*
22. Hobbs, D. *Op Cit.* P.22.
23. 'Early British Naval Aviation.' This undated paper was written by the then Air Commodore Samson after his retirement. Copy in IWM via G Bruce.
24. 'History of Naval Aeroplanes and Seaplanes,' by Colonel C.R. Samson, DSO. Stamped 30/12/18. TNA AIR1/724/76/2.
25. 'Developments at Eastchurch.' *The Aeroplane,* 17 October 1912. P.195
26. 'Messrs. Short Brothers New Aeroplane Works at Rochester,' in 'British Notes of the Week.' *Flight* 7 March 1914. P.251.
27. Report of the Committee of Imperial Defence Paper 139-B. At the 116th Meeting of the Committee of Imperial Defence on 25 April 1912, H.H. Asquith, the PM, stated that *he had approved this report provisionally in view of the urgency of the matter, and action upon it had been taken.* Copy via Gordon Bruce.
28. 'Memorandum on Naval and Military Aviation,' Command Paper 606T. Preface is dated 11 April 1912. RAFM AC TS/21/354. Copy via Gordon Bruce.
29. Rear Adm E.C.T. Troubridge, Chief of War Staff, to First Lord (Churchill). No date. Although signed by Troubridge, the handwritten memorandum is in Samson's handwriting. The memorandum was in reply to Troubridge asking Capt Murray Sueter and Samson for suggestions for the purchase of aircraft by the Admiralty on 23 January 1912. TNA AIR1/626/17/91.
30. Letter to author from G Bruce. 20.04.2020.

Chapter 2: Early Shorts Aircraft

Before anyone reads this, he should dismiss from his mind Aviation as it is to-day, otherwise he is apt to treat the efforts and experiences of my companions and myself as rather kindergarten; this is of course not correct as most of the time we were working completely in the dark without the mass of experimental data and laboratory knowledge that is available at the present day. Even to me, some of my early difficulties appear absurd at the present time.

Charles Rumsey Samson. *'Early British Naval Aviation.'*[1]

Short No.1 & No.2. Pusher Biplanes

Frank McClean ordered Horace Short to build him an aeroplane after he had flown with Wilbur Wright at Le Mans in 1908. Apparently, this was all he specified. Designed before the Wright brothers' contract was obtained, the S.1 superficially resembled the Wrights with landing skids and no wheels but the rigidly braced three-bay biplane wings had flexible tailing extensions at the outer bay for lateral control. It was a three-bay biplane with a front elevator, similar to those that featured on so many of the early aircraft, and a fixed fin. The engine drove twin pusher propellers by means of shafts and chains similar to the Wright method.

McClean fitted a 30-hp Nordenfelt engine from a car but it proved too heavy. It was not until the Wright type engine built by Bariquand & Marre arrived that the machine almost managed to get airborne. On the last attempt it stalled, crashed and was written off.

Specifications S.1: Span 40 ft; Length 24 ft 7 in. Area 576 ft^2.

The S.2 was designed by Horace Short for J.C.T. Moore-Brabazon for a machine to attempt to win the *Daily Mail's* prize of £1,000 for a circuit of one mile by a British aircraft flown by a British pilot. At the end of the wings were positioned 'balancers' for lateral control. The pilot sat on the edge of the lower wing to starboard of the centrally mounted engine. He operated the 'balancers' by foot pedals. A passenger could be carried on the opposite side of the engine.

The S.2 was a pusher biplane with three-bay wings and skid chassis. The machine had a forward elevator and rudder supported by the chassis that was extended forward.

As the British manufactured Green engine was not available it was fitted with a heavy Vivinus engine of Belgian make in order to enable trials to be conducted. After the Green was fitted, Moore-Brabazon won the prize on 30 October 1919.

Specifications S.2: Span 48 ft 4 in; Length 32 ft; Area 450 ft^2; Speed 45 mph.

Short-Wright Biplanes Nos. 1 to 6.

The Short brothers were impressed with the Wright Improved Model A that Wilbur Wright demonstrated in France in 1908, and they were contracted by the Wrights to build six machines for Wright customers in the UK. As there were no plans, Horace Short spent several days measuring and sketching the Flyer.

The first four Short-Wrights were similar to Wilbur's machine, the final two incorporated improvements. McClean ordered the third and while the others were waiting for their engines a Nordenfeldt engine was tried but the machine would not fly with it.

The six machines were manufactured for the following:

M Egerton (No. 4); C Grace (No. 5); F.K. McClean (No. 3); A Ogilvie (No. 2); C.S. Rolls (Nos. 1 & 6).

What is remarkable about these machines is that Ogilvie's machine was taken into the RNAS with Serial No. 1373 when it was purchased from the owner on his joining the RN in 1915. Contract No. CP66325/15 was raised to cover the purchase of the 35-hp Wright engined pusher biplane. It was at Eastchurch until September 1915 and was finally deleted on 13 January 1916.

Specifications Short-Wright:
Span 42 ft; Length 29 ft; Area 515 ft^2; Weight Empty 885 lb; Loaded 1,200 lb. Speed 50 mph.

The Short Brothers now gave up the Wright designs for the Farman-Sommer French biplane designs. According to Hugh Driver, this change was to lead directly to a turning point in the Short aircraft designs; *it was to lead directly to an influential series of original Short-Admiralty seaplanes.*[2] The Short designs that follow are grouped as to type so that the evolution of the machines can be followed, rather than just following the Short construction number (c/n) sequence.

At this point in the discussion of the early Short aircraft, it is convenient to list the missing Short 'S' numbers that were applied to orders that were not completed, other peoples' designs or were for spares, etc.

Short S.26. Short-Sommer Pusher Biplane

The S.26 was a variation of the Sommer Boxkite. This was the first machine that Shorts built after moving from Battersea to Eastchurch. Four of the type were built. Originally intended for George Cyril Colmore as the S.18, it was not completed and emerged as the S.26. It was fitted with a 40-hp Green engine, the single rudder was positioned below the tailplane and half-hoops below each lower tail boom formed the tail skid. It was flying by June

Right: Short No. 2 built for Moore-Brabazon in 1909.

Right: Short No. 2 in its final form with extended tail, 1909.

Right: Cecil Grace in a Short- Sommer biplane .The original caption indicates that this was the aircraft on which Grace was lost on 22 December 1910.– The Shorts Order Book shows that this was McClean's S33 lent to Grace for the Baron de Forest competition .

Short Job Nos. Shorts Aircraft and Other Tasks			
Job	**Type**	**Details of Order**	**Notes**
4*	Huntington-Dunne Triplane		Apr 1909
7*	Dunne D5 biplane	Blair-Atholl Syndicate.	Jun 1909
8*	Biplane glider	C.S. Rolls No.2, not completed.	Aug 1909
13	Biplane	M Egerton, twin propellers, crashed 11 June 1910.	Dec 1909
14	Spares for Ogilvie Wright		Dec 1909
15	Monoplane	G.C. Colmore, not completed.	Jan 1910
16	Spares, two Wright propellors	C.S. Grace	Jan 1910
17	Spares, two propellers	J.T.C. Moore-Brabazon	Jan 1910
18	Biplane – same as Job 13	G.C. Colmore, renumbered as S.26.	Jan 1910
19	Spares	A Ogilvie	Jan 1910
20	Biplane	As Job 13. Customer not identified.	Jan 1910
21	Spares & Launching rail & tower	C.S. Rolls.	Feb 1910
22	Propellers & other spares	C.S. Grace	Feb 1910
23	Propeller for Job No. 13	M Egerton	Feb 1910
	Small biplane	C.S. Rolls powered glider No. IV with 35-hp Green engine. No Job No. traced.	Mar 1910
24	Various tasks in fitter's shop.		Mar 1910
25	Repairs for C.S. Grace		

* Job No. deduced from Order Book. S.26 was introduction of c/n.
Source: Tabulation via G Bruce.

1910, but Colmore crashed it on 13 August, when he hit a fir plantation in Champel's Wood. In September it suffered another heavy landing at Colmore's hands.

It was then rebuilt with modifications. The repaired machine was acquired by McClean and listed as No. 1 in his private fleet of aeroplanes. Fitted with a Gnôme rotary engine, this was one of the machines that McClean made available for the Admiralty to train Navy pilots at Eastchurch. Surviving till April 1911, it had deteriorated to such an extent that it was used only for taxiing lessons. As Naval biplane No. 1 it was known as 'The Dud.'[3]

Short S.27. Tandem Twin

Built for Cecil S Grace it was another Sommer Boxkite variation. Powered by a 60-hp E.N.V. engine, it had rudders both above and below the tailplane. Known to be flying by June 1910. Various modifications were made as dictated by experience.

After Cecil Grace disappeared over the Channel McClean purchased S.27 and James Lindsay Travers converted it into the 'Tandem Twin' with the pilot seated between the two Gnôme rotary engines. It was known as the 'Gnôme Sandwich.' It retained its original wings and tail unit. The chassis had to be strengthened – Travers' undercarriages were known for their robustness. The Admiralty declined to purchase the machine, but McClean loaned it to the Naval Aviation School at Eastchurch where C.R. Samson crashed it on 11 March 1912, and it was apparently then written off. It received the Serial No. 4 in the first Admiralty system, and may have received the serial B4 in the second system.[4]

Wing Cmdr Stanley T Freeman recalled that the *S.27 was used in early experiments in bomb dropping. An experimental device was fitted under the lower centre section, carrying a 1-cwt weight that was released by a lever operated by the pilot. Samson tested it. The "missile" dropped cleanly and all the spectator were surprised that Samson continued to fly over the dropping area. When he finally landed, he was in a towering rage and the release lever was considerably bent. This was because he had expected the aircraft to jerk, rise, or give some similar indication of the release of the load, but it had not done so. Samson therefore thought that the load had failed to drop and he had made increasingly strenuous efforts to release it.*[5]

Specifications Short S.27:
Span 34 ft 2 in (later 46 ft 5 in wing extensions); Length 45 ft; Area 480 ft^2 (later 517 ft^2 wing extensions); Weight empty 1,800 lb; Loaded 2,100 lb. Speed 55 mph.

Above: Cecil Grace's No.2, as stencilled on the top part of the rudder with racing number '10' on the lower part. Bournemouth, 1910.

Above: Short S.27, possibly photographed at Margate. The fuel tank and seating arrangements have been revised.

Left: The modified Short S.38 with flotation chambers under the lower wings being man handled over the sea wall to ramps to carry it to the boat that will take it out to HMS *Hibernia*.

Short S.28

The S.28 was another pusher biplane similar to the S.26-S.27. It became the Moore-Brabazon No. 6. Purchased by McClean as his No. 5 after the 1910 Bournemouth Meeting where it was not flown. Made available by McClean for naval training in 1911.

A pusher, it was given a nacelle for the pilot and pupil. On 1 May 1911, it crashed from 80 feet and was badly wrecked. Samson recommended that the machine be replaced by a new one rather than repairing the wreck. The engine and nacelle were incorporated in its replacement, the S.38.

Short S.30 & S.31

Pusher biplanes. Not built.

Short S.32 – Short-Sommer No. 2 Biplane

In 1912 an offer to train members of the London Balloon Company and the Territorial Force at Eastchurch was announced. Again, it was McClean who provided the S.32 and paid for the maintenance of the aircraft.[6]

The S.32, McClean's No. 8, was rebuilt with a 70 ft 6 in span as the S.58 for a proposed journey up the Nile River, and was featured in the 7 June 1913, issue of *Flight* under the title of the '70-H.P. Short Biplane.' The article commences with a salute to Messrs Short Brothers who *execute so much confidential work for the Admiralty,* however its statement that McClean was favourably impresses with its qualities as a land machine is far from the truth. The machine proved to be very slow and could make no headway against any appreciable wind. The tandem nacelle also did not allow for the storage of equipment that would be necessary for the proposed expedition. This machine was replaced by the S.80 that became the Nile Expedition machine.

Specifications S.32 rebuilt:
Span 70 ft 6 in; Length 35 ft 6 in; Ares 725 ft^2; Weight Loaded 1,540 lb; Speed 38 mph.

Short S.33

Cecil S Grace entered this machine in the Baron de Forest Contest for the first British machine with a British pilot to cross the Channel. Grace was lost in the Channel on this machine on 22 December 1910, when returning to the UK.

Short S.34 'Long Range' Pusher Biplane. RNAS No. 1/B.1/T.1.

This biplane was one of those made available by Frank McClean to train Navy aviators. In June 1911, the nacelle was removed from both the S.34 and S.38. Purchased by the Admiralty in late 1911, it was originally Navy biplane No. 1, then B1, then T1. It had made a world record cross-country flight on 16 August 1911, of 4 hours 13 minutes.

The Shorts' catalogue described the School Biplane Type S 34 as *specially designed to withstand the rough wear and tear of school work, whilst giving the greatest amount of safety to pupils.*

It is a biplane of the engine behind type, fitted with either a 50 H.P. or 80 H.P. engine.[7] *The teacher and pupil seats are arranged side by side, and a system of dual control is provided which enables the pupil to follow every movement of the instructor in controlling he machine. In this manner, the pupil from the start occupies his position that he would occupy in actually piloting the machine himself, and gives him full control of warping, elevating, and steering himself, an advantage which will be readily recognised by those who have been taught by the old method, which generally left the pupil unacquainted with the steering until he began his pilot flights.*

In general construction the machine is equal in quality and finish to the higher speed machines, the same material being used.

The planes are covered with SHORTS' improved fabric, and the usual engine speed indicator, throttle, sight-feed lubricators, etc., are provided. The machine is in general favour amongst naval and military officers for school work.

Jane's All the World's Aircraft 1913 described the S.34 as solely designed for school work. Seats side by side.

Specifications Short S 34

Engine	50-hp Gnôme		70-hp Gnôme	
Span	46 ft 5 in	14.18 m	46 ft 5 in	14.18 m
Length	42 ft 1 in	12.85 m	42 ft 1 in	12.85 m
Height	13 ft 11 in	4.27 m	13 ft 11 in	4.27 m
Weight	1,100 lb	500 kg	1,150 lb	523 kg
Speed max	39 mph	64 km	48 mph	77 km
Speed min	34 mph	55 km	38 mph	61
Fuel Capacity	4 gal	..	5 gal	..

Source: *Catalogue of Naval and Military Aeroplanes. Short Brothers*, Catalogue A. No date.

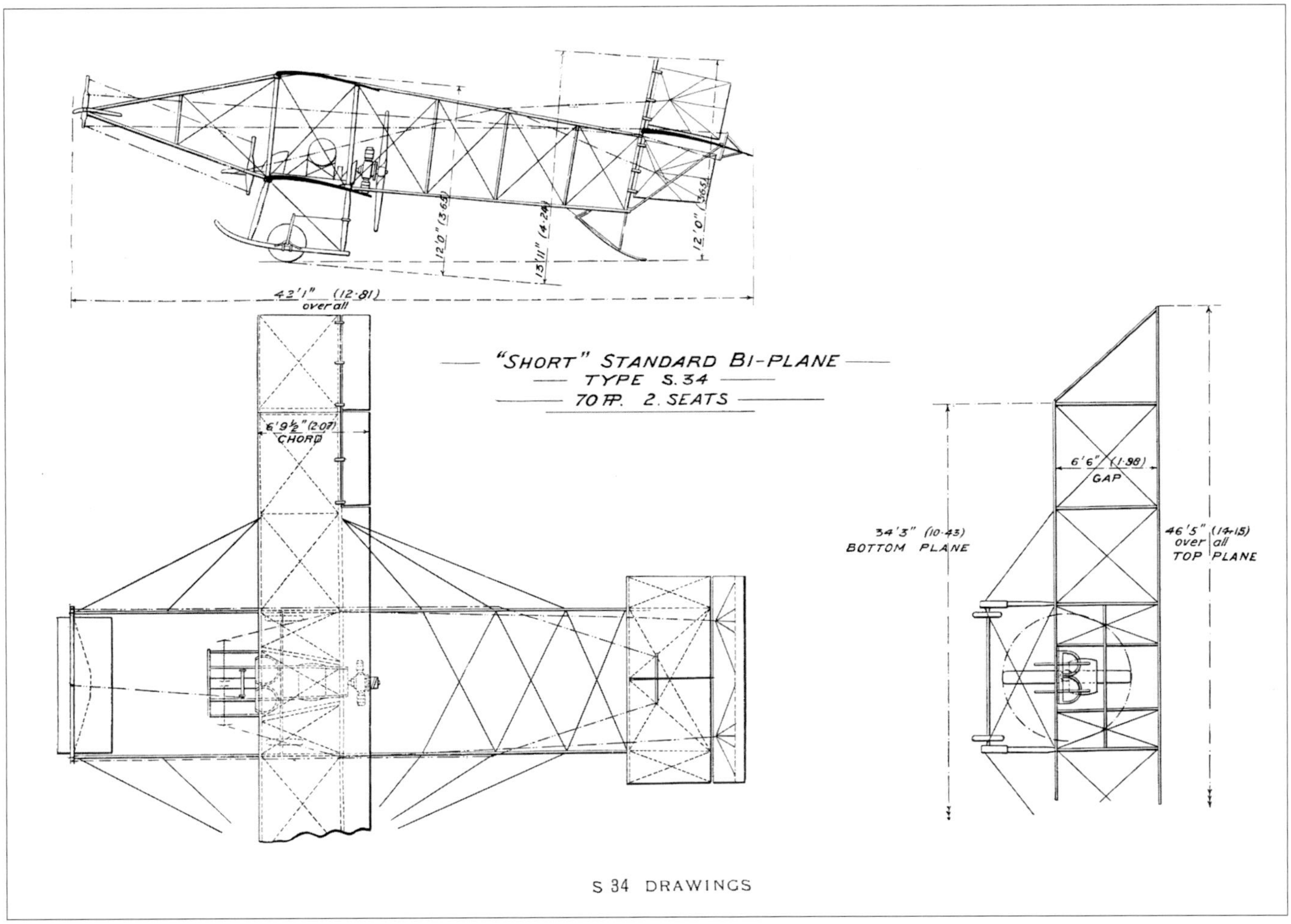

Above: Drawing of the S.34 from the Shorts' catalogue.

Above: Lt Malone and G.S. Trewin at Margate in the S.34, 30 May 1912. This machine was later built to S.38 standard as the S.86.

Samson flew B1 carrying a 98-lb dummy bomb and dropped it from 80 feet, and *felt no bump at all.*[8] This statement highlights how everything that the RNAS was doing at this time was experimental. They had no predecessors to help them develop naval aviation. *The Aeroplane* reported that on Monday, 15 December 1913, the Lt Clarke-Hall was piloted over Eastchurch aerodrome by Sub-Lt Peirse, RNR, in the *Short biplane (S 34),* dropping bombs from 2,000 feet. The bombs were carried in a box beneath the lower plane and operated by a pedal release. The following Wednesday, Clarke-Hall *flew to the Isle of Grain and there dropped bombs, getting one bomb right in the net, using Short 34 as before.*[9]

Short S.36

The S.36 was the first tractor biplane of a *series of sensible tractor biplanes* that Short Brothers produced in late 1911.[10] This machine had been designed for C.S. Grace but the plans were filed on his death. Frank McClean was shown the plans in the Autumn of 1911 and was so impressed that he ordered one built as the McClean No. 12. Tested by Cdr Samson and

Specifications Short S 36

Source	1	2
Span	46 ft 5 in	45 ft 5 in
Length	35 ft 6 in	35 ft 6 in
Areas in ft²	515	515
Empty	850 lb	850 lb
Loaded	1,300 lb	1,300 lb
Max Speed	60 mph	60 mph
Endurance	5 hrs	
Motor	70-hp Gnôme	

Source: 1. M.H. Goodall, & A.E. Tagg data.
2. C.H. Barnes data.

Capt Gerrard in early 1912, the rear fuselage was originally devoid of fabric but later covered in. McClean lent this 70-hp Gnôme machine to the Navy School at Eastchurch. This aircraft led to the S.45.[11]

The Short S.38 Type

The S.38 would be built in series but, as with so many of the early Shorts, there were marked variations between individual S.38 aircraft. The product that Shorts advertised and sold as their S 38 was very much changed from the original S.38 biplane. In its best-known form, the S.38 had a long nacelle with a forward elevator perched on its prow.

Fourteen S.38 Types were built, the majority for the Navy although the Army purchased one (S.62 Serial No. 446) for the Central Flying School (CFS).

Grahame Donald flew a variety of aircraft at Eastchurch, and the Short he describes was most likely a Short 38 Type. *The old Short biplane was a very nice 'plane. It was a pusher; rotary engine 70 horse power Gnome. Slightly bigger than 50 horse. And a little short fuselage or nacelle sticking out in front of the wings – it sat there. Very small elevator on the nose of the nacelle; it didn't do very much, but was quite nice to have there. You could keep your eye on it for a sideways balance and 'fore and 'aft. Two tail booms with a good tail fin elevator and rudder on it. And a very pleasant plane to fly. She was a little bit tail-heavy, but not unduly so. But you had to watch her. If the engine happened to cut she was inclined to drop her tail – but you got her nose down quickly if the engine stopped. … The most amusing Short I flew was Naval No. 1. It was interesting because it had side-by-side seats with a wide cockpit in front. Side-by-side seats fitted with dual. …And it was very pleasant to have flown it.*[12]

Short S.38 Sommer Pusher Biplane – 'Improved S.27'. RNAS No. 2.

The Short S.38 was a pusher biplane described as a Short-Sommer type. It was the replacement for the S.28 and incorporated parts and the engine of that pusher biplane trainer. It appears that the machine was ordered by the Admiralty, but as the Navy was responsible for accidents to McClean's aeroplanes, it was recorded as *F. McClean (Navy)* in the Shorts' Order Book. The S.38 incorporated the nacelle of the S.28, had ailerons on both upper and lower wings and king post braced upper wing extensions. The S.38 first flew on 24 May 1911, with Samson at the controls. On 19 August, Samson flew the machine for 4 hours 58½ minutes for a new British duration record. In June the nacelle was removed. The Admiralty purchased the machine in the latter half of 1911 for £600. It received the Navy serials 2/B2/T2/2.

Oswald Short wrote that the development of a seaplane was *approached with great caution by the experimenters of the day. The effect of alighting on the water at high speed was an unknown quantity. It was assumed that in any case the resistance of the water to the floats would be very great, and it appeared highly probable that the high centre of gravity of the machine in relation to the point of contact between the floats and the water would create a couple causing the machine to turn over in spite of the force which might be brought to bear with the tail plane and elevator to counteract this tendency.*[13]

It is a remarkable fact that possibly the first machine to land on water in this country did so by means of rubber bags inflated with air.[14] Lt Arthur Longmore and Oswald Short devised a method of fitting air-bags to the S.38. Oswald learned from his ballooning days of *the enormous strength of rubberised fabric, and I began building a very light bag which could be attached to the skids.* In order to test the strength of the bag, he filled one with water and dropped it from *the keel of the machine which I had suspended in mid-air. That bag never burst, and I remember Lieut. Longmore, the pilot, exclaiming "That's good enough for me."*[15]

In addition to the wheels, two large torpedo-shaped

Right & Below: The 80-hp Gnôme powered S.36 as originally built with open rear fuselage, completed for Frank McClean, November 1911.

Right: The caption McClean's Aeroplane Mark-Cross, July 2nd 1912, says it all. The biplane has an early Shorts' logo stencilled on the lower part of the rudder. Shorts were soon to realise the benefits of displaying their name more prominently on their aeroplanes. The rear fuselage had been covered in by May 1912.

Above: Frank McClean and his sister in the S.36 at Eastchurch, 1912.

air-bags were fitted under the lower wing, attached to the undercarriage struts, with a third air-bag under the tail. Longmore successfully alighted on the Medway on 1 December 1911, in this aircraft. As the machine could not take off from water, it was brought to shore by a Naval party and flown back to Eastchurch from land.

Lt C.R. Samson, who had become the CO of Eastchurch in 1911, *designed and had built in Chatham Dockyard a platform for flying off the deck of ships. ... This platform, which was a double trackway, was my sole personal design. It proved successful, and although the Dockyards had made it much heavier than I intended, it was quite a workable contrivance being designed to take to pieces.* This was mounted over the fore gun turret of the pre-Dreadnought HMS *Africa* that was in the nearby Sheerness Dockyard.[16] On 10 January 1912[17], he flew over to Sheerness Harbour near Eastchurch in his air-bag equipped Short S.38. *From this platform he made a successful ascent exactly half-an-hour after the aeroplane had been brought alongside. When the engine was started up the biplane was held back by blue-jackets, but having found everything was in order, Lieut. Samson gave the signal to "let go," and the machine ran down the staging, and after passing the bows rose steadily to a height about 100 feet, amid cheers from* Africa *and the warships in the vicinity. The pilot encircled the battleship at an altitude of several feet below her topmast, and then rose to a height of 300 feet and flew back to Eastchurch.*[18]

In his post-war recollections Samson wrote that the *trackway looked very short and I didn't feel at all confident that I would be able to get into the air. The mooring buoy looked 'grim' as if awaiting my impact. Unfortunately owing to the strength of the tide being more powerful than the wind the ship was not lying head to wind and tugs were signalled for. I realised however, that the various high officials on the bridge wouldn't like a wait so I said to the Commander, "I won't wait for the tugs". I pulled the quick release and started to toddle, and 'toddle' actually describes how it felt to me, along towards the bows. I found a little difficulty at first in keeping on to the trackway but fortunately no mishap occurred. As I neared the end I got my tail well up and I didn't attempt to get her nose up until I was at the very end of the run. To my delight I found myself flying, but I only missed that mooring buoy by about three inches.*[19]

This was the first flight from a British warship. On 2 May, he repeated this feat from HMS *Hibernia*, to which the runway from *Africa* had been transferred. *I flew off the trackway erected on the "Hibernia" whilst she was steaming at 10 ½ knots.*[20] *I was using the same aeroplane, but we had fitted a 70 h.p. Gnome in place of the original 50 h,p. This was the first flight ever made from a ship steaming. I found no difficulty at all and left the trackway after a run of 60 feet.*[21]

Samson described the S.38 as *very nasty*.[22] This may have been due to the engine falling off in the air on 15 April 1912[23], while he was flying the machine, rather than its habits while taking off from the ship's deck.

On 3 July, Samson landed the S.38 alongside HMS *London* that had been equipped with a flying-off platform. He was hoisted aboard and proceeded to the Portsmouth Naval Review. The following day with *London* steaming at 12 knots, Lt C.J. L'Estrange Malone took-off into a 20 knot wind and made a successful flight over 19 miles to land at Eastney Barracks. It was said - probably apocryphal – that Malone took off without any forward run, beating the Sea Harrier by some 80 years. S.38 was wrecked on 9 July 1912, while being hoisted aboard its lighter in a rough sea.

'Converted,' it emerged in late Farman style with a nacelle. It was flying again by 29 August 1912. C.G. Grey of *The Aeroplane* was at Eastchurch in October where he was

Specifications Short S.38

Source	1	1	2	3	3s
Span upper	32 ft	52 ft	52 ft	52 ft 0 in	52 ft 0 in
Span lower			34 ft		
Length	35½ ft	35½	35 ft 6 in	35 ft 6 in	35 ft 6 in
Height			517	11 ft 6 in	11 ft 6 in
Empty	950 lb	1,050 lb	1,100 lb	950 lb	1,050 lb
Loaded			1,540 lb		
Speed in mph			$39^1 - 45^2$		
Max	42	58		42	58
Min	35	39		35	39
Endurance	4 hrs	5 hrs		4 hrs	5 hrs
Engine	50-hp Gnôme	80-hp Gnôme	50-hp - 70-hp Gnôme	50-hp Gnôme	80-hp Gnôme

Source: 1. *Jane's All the World's Aircraft 1913*.
2. C.H. Barnes data.
3. Shorts *Catalogue of Naval and Military Aeroplanes*.

Left & Above: The air bags under the wings of the S.38 can be clearly seen while it is being hoisted aboard HMS *Hibernia* at Weymouth in May 1912.

Above: S.38 safely aboard HMS *Hibernia* on the ramp built over the superstructure of the ship.

Above & Below: The ramp over HMS *Hibernia's* guns was only a temporary feature and the whole exercise was an experiment to determine whether an aeroplane could take-off from a ship.

Left & Above: Cdr Samson taking off from HMS *Hibernia*, May 1912.

Right: Short S.38 behind the S.41 on HMS *Hibernia*. For the Naval Review at Weymouth in May 1912.

Below & Below Right: Taking the S.38 out to HMS *Africa*.

Above: The S.38 landing after its take-off from HMS *Africa*.

Right: The original S.38 was wrecked while being hoisted aboard HMS *London* on 9 July 1912. It is shown after its reconstruction at Westgate in January 1913.

Left: The rebuilt No. 2 under the command of Flt Sub-Lt J.P. Wilson at Eastchurch, September 1914.

Left & Above: Serial No. 65 (c/n S.75) taking-off.

Top Left & Right: Serial No. 28 (c/n S.55) a Type S.38 pusher, force landed at Rackington, Kent, where these photographs were taken.

Right: Serial No. 8434 was the first of a batch of six S.38 trainer biplanes built by Norman Thompson Flight Co Ltd with 80-hp Gnôme engines.

Below: Serial No. 8437 was from the same batch. No. 8347 is recorded as having a modified undercarriage. Delivered to Chingford on 27 January 1916, it was dismantled on 8 April and deleted on 22 April 1916.

Above: Short S.78 was the S.39 (Serial No. 3) rebuilt into the S.38 nacelle pusher layout. Photographed at Eastchurch. Flown by 4 June 1913. After the outbreak of war, it served on the Eastchurch patrol, having been allotted a Maxim machine gun in August. It was finally deleted on 26 April 1916.

Above: Short S.78 taking off. The Serial No. 3 is carried on the rudder.

Above: M Edgerton's S.59 before it was impressed after war was declared. The Shorts' logo is prominently displayed on the rudder with the c/n 'S 59' at the lower edge. (Short Brothers photo)

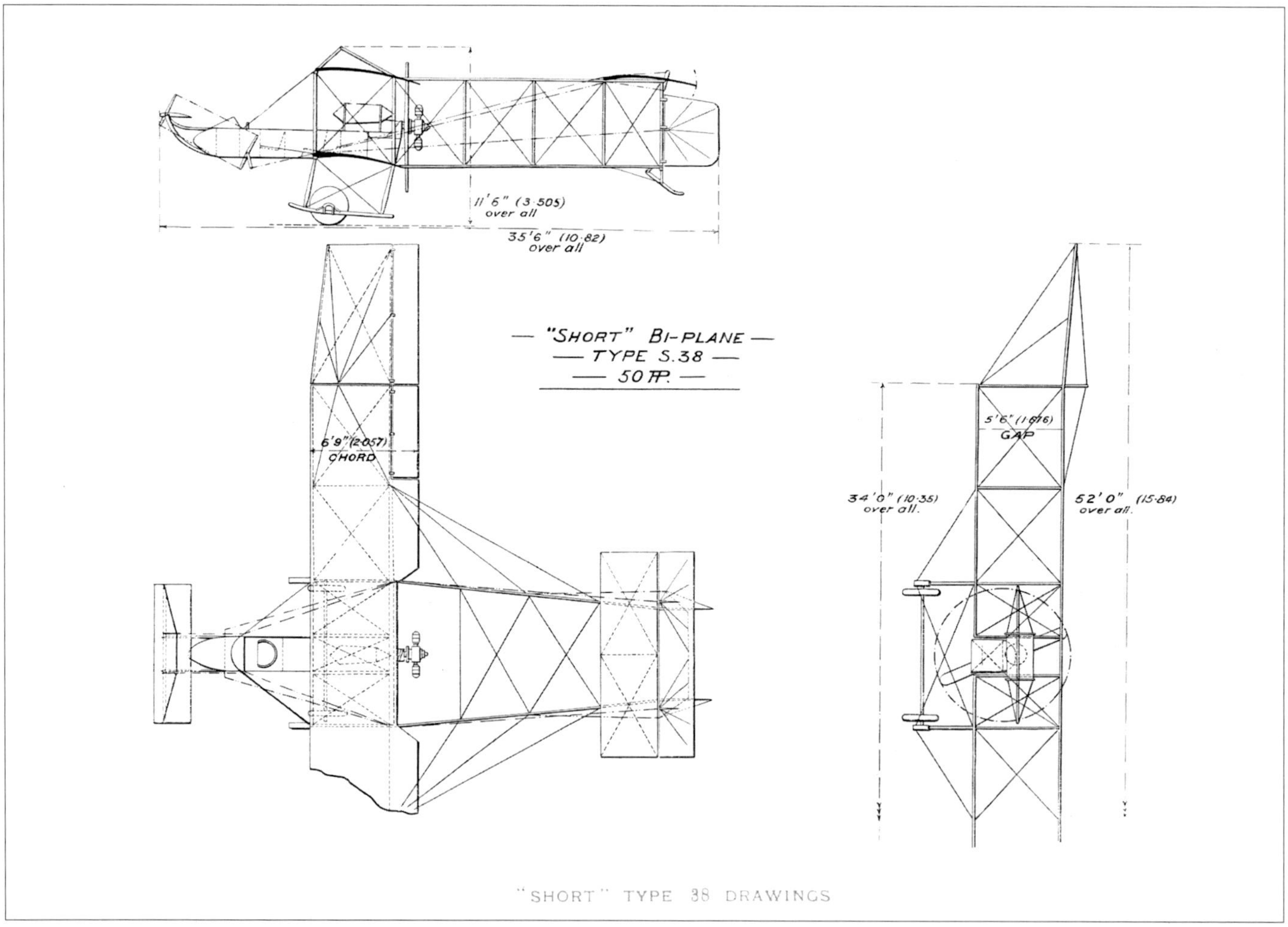

Above: Short S.38 Drawings from Shorts Catalogue.

told that Lt John Wilfrid Seddon, RN, was up in a Short biplane. Seddon was *he who broke his leg early this year by attempting to land another biplane inside a shed which had its door closed.* When he landed Seddon's machine was revealed to be *the ancient "Short 38" rebuilt. In its new form the extended sections of the upper plane have a pronounced dihedral angle, and taper sharply backwards. The gap between the planes has been reduced. The chassis is lower. The engine and its bearers are considerably above the lower plane, and are fixed to a raised body or fuselage, in which the pilot sits a good way in front of the leading edge.* The passenger sat behind the pilot where he had plenty of room to stow maps, drawing gear and instruments. *The front elevator is very small and practically flat, and it is placed only a few feet in front of the pilot; in fact it seems to be there to keep him company more than for any other reason. As a matter of fact, it has quite a lot of effect, for the machine is very quick on her controls, though practically automatically stable when once in the air.* The machine had a 50-hp Gnôme engine and being a large machine, *and consequently she is one of the slowest machines in the Service, for her power goes in lifting efficiency and stability.*[24]

It performed one of the earliest night flights by an aircraft on 7 September when flown by Lt Wilfred Parke, RN.[25] Used as a trainer at Eastchurch it underwent more repairs and reconstruction. Its final crash that ended its career occurred on 28 January 1915. The S.38 had a long and eventful life for a 1911 machine being finally written off in February 1915. AM Sir Ralph Sorley flew S.38 landplanes and noted that they were very lightly loaded per square foot of wing area.[26]

Jane's All the World's Aircraft 1913 described the S.38 Military Nacelle Biplane as *specially designed for reconnaissance. Tandem seats, pilot in front. An extra passenger can be accompanied.*

Short S.58 – Side-by-side School Pusher Biplane. RNAS No. 904

As noted above, the S.58 was S.40 rebuilt with an increased wingspan. The S.58 featured a side-by-side nacelle and was *very much like the rebuilt S.38*. The forward elevator was carried very close to the nacelle. A 70-hp Gnôme was installed in pusher configuration. Three-bay wings with a span of 70 ft 6 in replaced the original two-bay wings. The large overhang of the upper wing was braced by diagonal struts from the lower wing. The high-set tailplane was supported by booms and had twin rudders underneath.[27]

Impressed on outbreak of the war under Contract CP65176/14, it was delivered to Eastchurch in August 1918. It was fitted with standard 52 ft span wings and performed bombing experiments at Grain in February 1915. It was deleted by the end of June1915.

Short S.78 – S.38 Nacelle Type Pusher biplane. RNAS T3/No. 3

This aircraft again demonstrates the problems with recording a straight narrative of the early Short aircraft, being a rebuild of the S.39 into an S.38 type.

Unsuccessful as a twin, in December 1912, the S.39, No. 3, was returned to the makers for overhaul but was completely rebuilt under Contract C.P.31105/13/X.2087 as S.78. Writing in his 14 August 1913, issue of *The Aeroplane*, C.G. Grey noted that at Eastchurch was *one of the prettiest biplanes I have met. This bears the Naval number "3," and is apparently built out of some prehistoric Short, probably about "S.28," or so. Its general arrangement suggests the Henry Farman, but aerodynamically it is quite different, particularly the wing curve. Its speed is about 63 m.p.h. and its climb is alarming … after seeing a Navy pilot take it up like a rocket, I believe it.*[28] Richard Bell Davies recalled that it was common practice to send the machines back to the manufacturer to be repaired and for a new machine to emerge from the factory.[29]

Carrying the same official serial, No. 3, this tandem seat pusher biplane appeared in June 1913, with an 80-hp Gnôme engine. By 6 August it had a 70-hp Gnôme. On 23 October 1913, Samson took Winston Churchill, the First Lord of the Admiralty, over to Grain Island in No. 3. He recorded the flight took *12 mins 9' – 1200 ft.*[30] *The Aeroplane* recorded that Churchill had flown in practically every type of machine that the Navy possessed. His thoroughness in acquiring knowledge, and his readiness to share the risks taken by the Service, together with the excellent effect produced by thus showing his confidence in the Air Department's choice of machines and pilots, may help to an understanding of the high esteem in which the Navy holds its political chief.[31]

No. 3 participated in the 1913 Army Manoeuvres. *The Aeroplane* reported, that on Thursday 18 December 1913, *Comdr. Samson, R.N., was making Short 3 turn round and round in small circles, engine running, with nose up so that it was literally turning on a wing tip without getting any nearer the ground.*[32] After the outbreak of war, it was allotted a Maxim gun and served on the Eastchurch patrol. Wing Cdr Samson had flown it several times at night pre-war and was used to the aircraft, but on 31 October 1914, the propeller burst causing a major crash. It spent a few weeks at Dunkerque where it was repaired. Returned to Eastchurch it remained there until deleted on 26 April 1916.

S.55 & S.61 – S.38 Type Nacelle biplanes. RNAS Nos. 28 & 34

Two Short S.38 Type biplanes, No. 28 (S.55) and No. 34 (S.61) were purchased separately in 1912 and 1913. No. 28 was a 50-hp Gnôme powered nacelle type pusher biplane. The machine was noted as having a 70-hp Gnôme engine by *The Aeroplane* when it reported that the new Short piloted by Lt Parke, RN, with two passengers, rose to 1,000 feet in

9 minutes. In a later issue, the journal corrected this, noting that at the time of the flight it had a 50-hp Gnôme, the 70-hp being installed later. [33] *The Aeroplane* reported that *the new Short S.55, practically a sister machine to S.38,* had conducted its official trials at Eastchurch, carrying Lt Parke with E.A.R. James as passenger, on 4 November 1913.[34] No. 28 was used in early bomb dropping experiments from May 1913. For the Week Ending 24 May, Carpenter Jas V Collins recorded on his 'Work Done' sheet - *Replacing Bomb dropping gear to No. 28.*[35] In August 1914, it was allocated to Great Yarmouth but was blown over by a gale on arrival, and the wreckage was transported back to Eastchurch and repaired. After serving at Eastbourne in March 1915, it was rendered unserviceable and was out of action until November. The following January it was again requiring repairs. It was deleted in March 1916.

Arnold Hugh Sandwell was trained at Eastchurch. He had enlisted in May 1915, after his return from Canada. *The RNAS flying school at Eastchurch was run on humane and commonsense lines.* The CO was Sqn Cdr Philip Shepherd, a real Navy man who had served in the Persian Gulf, and *could swear in five or six languages including Arabic and Hindustani.* Eastchurch aerodrome was surrounded by dykes, and one that bisected the landing field was bridged over. *The machines were of heterogeneous origin and were there either because they were unsuitable for war service or had been retired on account of old age and/or obsolescence. … There was a Short pusher with visiting-card elevators in front, on which I had my first tuition. Its official number was 28. A similar machine numbered 3 had at least been rebuilt. Heaven knows how many pilots had learned to fly (if it* was *flying?) on these machines.* After 68 minutes dual and 43 minutes in an Avro, Sandwell went solo on the Short. Before he could build up any time on No. 28 she was crashed by another pupil and he switched to the 80-hp Caudron tractor.[36]

No. 34 was also a separately purchased Type S.38 with an 80-hp Gnôme engine. At Eastchurch by May 1913, it was used for bomb dropping development. At Grain by September 1914, it underwent further armament trials and was fitted with dual controls. Deleted on 7 February 1915.

S.66, S.67, S.75 & S.77 – Short S.38 Nacelle pusher Type Biplanes. RNAS Nos. 62 to 66

Under Contract C.P. 36191/13 five Short S.38 pusher biplanes were ordered by the Admiralty and received serials Nos. 62 to 66 (Short S.66, S.67, S.76, S.75 and S.77 in sequence).

Flight reported that a new biplane built by Short Brothers for the Admiralty *has been tested by Mr Gordon Bell. This machine is designed to carry a quick-firing gun, which together with the operator will be accommodated in the forward part of the* nacelle *of the machine.* A weight equivalent to the gun and gunner was carried for the test. *The machine is of the usual type supplied to the services with the exception of the provision made for the gun.*[37] This was the S.77.

Serial No. 66 (S.77) was a rebuild of the S.28 Sommer type Boxkite, and with an 80-hp Gnôme, was first flown on 24 September 1913, and remained in service until April 1914.[38]

This would have been the Short *gun-machine* that *The Aeroplane* reported as being tested by Gordon Bell at Eastchurch on Wednesday 22 October 1913. Also, the *S.38 type, 80 Gnome, but without font elevator, passenger in front, with a small Maxim* that *The Aeroplane* reported at Eastchurch on Thursday 18 December 1913. The new Short gun-carrier was piloted by Lt Clarke-Hall, RN, with Sub-Lt Marix, RNR. No. 66 had been involved in experiments with Vickers and Maxim guns and wireless equipment. Samson recorded in his Log Book for 1 April 1914, that at 1050 he had taken *No. 66 (Gun Machine) carrying Lt Clark Hall RN firing Maxim gun at shadow thrown by No 66 on the water height 1500 ft good shooting. 40 mins 36'.* This was repeated on the 22nd that month.[39] In June 1914, it has a 50-hp Gnôme installed. By August it was on the Eastchurch Patrol. It served at Eastchurch until its deletion in April 1915.

Ralph Sorley flew the mix of aircraft at Eastchurch when he was a Flt Sub Lt in the RNAS. He flew S.38 trainers Nos. 62, 63 and 65. He took his 'ticket' – Royal Aero Club Certificate No. 1089 - on No. 63 on 12 February. He had made his first solo on the type on 5 January for 14 minutes. He remembered that the Short pusher biplanes of 50 and 70-hp Gnôme engines were very lightly loaded with their wide wingspan. *They also had ailerons which hung down vertically until sufficient speed had been attained, when they became effective. They also had a front elevator attached to the nose of the nacelle, but which had little or no effect as such; but it was a big help in lining up the horizon. They were light on all controls and very pleasant to fly, but with such light loading that they were liable to bounce into the air at take-off or on landing and one had to be very quick to 'blip' the engine to prevent a stall.*[40] Richard Bell Davies recorded that the Short landplane was called the 'Short bicycle.'

In the 15 July 1915, issue of *The Aeroplane*, it was reported that No. 65 had been fitted with wheels and floats.[41]

Short S.62 – S.38 Type Tractor biplane. RFC No. 446

S.62 was the only S.38 Type to go to the Military Wing of the RFC. Completed in early 1913, on Saturday 22 March, Gordon Bell, accompanied by his mechanic, flew the biplane to Hendon from Short Brothers factory at Eastchurch. When landed, a violent gust of wind turned the aircraft over onto its back. So violent was the gust that extensive damage was done to the aircraft. After it was rebuilt it was acquired by the RFC. Officially allotted to the Military Wing on 3 July 1913, as serial No. 446, it went to Upavon and was used by the CFS.

No. 446's flying career was brief, ending on 3 October

1913. Maj George C Merrick was descending for a landing when the machine dived steeply tossing the pilot out and crashing. Merrick was killed. Shorts had supplied safety belts with the machine but Merrick had not been wearing his and it was thought he had fallen onto the controls, steepening the descent so quickly and steeply that he had been thrown out of the cockpit. The Short had righted itself but turned over on striking the ground and was not repaired but struck off.

Jarrett suggests that No. 446 was S.63 and that S.62 was not delivered to the Army, but may have supplied parts to the modification of S.58.[42]

Specifications:
Span 52 ft; Length 36 ft; Wing area 500 ft^2; Weight Empty 1,050 lb; Loaded 1,500 lb; Speed 48 mph.[43]

Short S.89 – 'Sociable' Tractor Biplane. RNAS No. 152

This version of the S.38 had the crew sitting side by side, hence the term 'Sociable.' Accepted at the Shorts' Eastchurch factory on 9 June 1914, it was on the Yarmouth War Flight in July undertaking coastal patrols. With B Flight, Eastchurch in November that year. The 80-hp Gnôme engine was removed at Eastchurch by February 1916. It survived to be deleted on 27 April 1916, and it is presumed it served as a ground instructional airframe in the interim.

Short S.43 & S.44 – Pusher School Biplane RFC Nos. 401 & 402

Frank McClean offered flying training to members of the London Balloon Company, RE, TF, using his S.32 and similar S.33 pusher biplanes. Excellent trainers, two were ordered by the War Office (S.43 and S.44). They were in use in the second half of 1912. When the RFC serial system came into force in August 1912, they were numbered 401 and 402 respectively. They were in use up to November 1913, but their final disposition has not been discovered to date.

The Aeroplane for 4 July 1912, P.10, reported that a new hydro-aeroplane S.43 has been delivered to Naval Wing of RFC at Eastchurch by Short Bros. Tested by Samson over Sheerness Harbour with a passenger made 62 mph in a high wind. Fitted with wireless and carries 6 hours fuel. It will take part in naval manoeuvres in August.

S.43 was not a hydro-aeroplane, and it is difficult to determine which machine is being reported. It is thought that this machine could be either the S.45 or S.41. Both were initially flown as landplanes; however, *The Aeroplane* article

Left: The S.32 in 1911. Built for C.G. Grace, it was acquired by J.T. McClean.

Right: Short S.40 floatplane near Westminster.

Short S.38 Serial Allocation			
Serial Nos.	**Manufacturer**	**Contract No.**	**Notes**
401 - 402	Short Brothers		S.43 & S.44.
28 & 34			S.55 & S.61.
62-65		CP36191/13	S.66, S.67, S.76, S.75, S.77 respectively. No. 65 had both wheel & float undercarriages.
1580 – 1591	Supermarine	CP50249/15	Tender of £11,326 accepted W/E 1 July 1915. Delivered from November 1915.
3143 - 3148	White & Thompson	CP59218/15 & CP75962/15. (Total £11,400 for both.)	Tender accepted W/E 8 July 1915. Delivered from November 1915.
8434 – 8439	Norman Thompson Flight Co Ltd		Tender accepted W/E 8 July 1915. Delivered from December 1915.
8530 – 8541	White & Thompson	CP77567/15	Delivered from March 19156.

states it was a new hydro-aeroplane when delivered, further confusing the identity of this machine.

Specifications:
Span 45 ft 5 in; Length 42 ft 1 in, Wing Area 517 ft^2; Empty weight 1,100 lb; Loaded 1,540 lb. Speed 45 mph.

Licenced Production of the S.38

The 'production' model of the S.38 was used as a trainer. This had the small forward elevator, double rudders, and improved undercarriage with wheels and skids. As Short Bros were tied up with seaplane construction, contracts were given to other firms. White & Thompson Ltd, (later the Norman Thompson Ltd), of Bognor, built 24; while Pemberton-Billing Ltd (later Supermarine), of Southampton, also built 12 of the type.

A report on the Supermarine Ltd works at Southampton in September 1915, noted that this *firm's Works do not appear to be sufficiently large for the amount of work they have in hand, but it is understood they propose enlarging them in the near future.*

The question of putting the fabric on to the S.38 machines this firm is building for the Air Department was carefully gone into and the correct method explained in which this should be done, as owing to the peculiar construction of the main planes it is a matter of some difficulty to get the fabric on squarely and sufficiently taut without distorting the wings.

The firm are experiencing difficulties in connection with the wiring of the extension on these machines and it is proposed to modify the present system of using the three-strand cables and substituting cable for the main flying wires and piano wire for the two intermediate flying wires on this part of the machine. In the original machine as built by Messrs. Short only piano wires were employed, so that this alteration is merely reverting to the original design.

The firm's general workmanship appears to be good.[44]

The same week, an inspection of White & Thompson took place. This found that the workshops were very congested due to the large number of S.38 aircraft that were practically ready for erection *and only awaiting minor components, the raw material for which had not yet been obtained.* The firm had been given assistance with this and it was hoped that the first machine would be ready in a month's time.[45]

The type served at Great Yarmouth on coastal patrols, and for training at Chingford and Eastchurch.

One typical incident took place at Eastbourne. This aerodrome was established by the Eastbourne Aviation Company in 1911 and in 1913, it was expanded to incorporate a seaplane base. The Company operated their own flying school and also constructed a number of aeroplanes of their own design pre-war. The Company's facilities were taken over by the Admiralty on the outbreak of war as an RNAS training station. The Company continued to manufacture aircraft during the war with contracts for B.E.2c, Maurice Farman S.11 and Avro 504 trainers.

On 19 January 1916, WO2 Percival Victor Fraser, an Australian, was instructing Probationary Flt Lt Gordon Ezra Duke at Eastbourne on Short S.38 No. 3148. After take-off the biplane was seen to fly in an irregular manner and then nose-dive into the ground near a brickfield close to Hamden Park. Both occupants were killed in the crash. The cause of the incident was thought to have been a control cable that pulled off its control pulley.[46]

A few of the type that were in RNAS service were used for patrols during the early days of the war. No. 8438 from Felixstowe, with a crew of Flt Sub Lt C.J. Galpin and AM Pinn, undertook an anti-Zeppelin patrol on the night of 24/25 April 1916. Short 184 No. 8351 from Felixstowe and Short 827 No. 3105 from Yarmouth also took part in the hunt for the intruders. The raid was combined with a reconnaissance planned to assist the warships bombarding Lowestoft.[47]

The majority of the type saw out their service lives as trainers.

Short S.45. Tractor Biplane. RNAS No. T5/5

The Admiralty ordered a similar machine to McClean's S.36 that was built in the spring of 1912, and delivered to the Navy's school at Eastchurch on 23 May 1912. It was different from McClean's machine only in having shorter wings. The upper wing overhang was supported at their extreme tips on the Admiralty's machine. Originally a landplane, by 2 July 1912, it was fitted with a large central float. Wing Cdr Stanley T Freeman recalled that T5 was very difficult to handle on the water as it did not readily right itself after heeling over when turning. Freeman was sent out onto the wing to help right the aircraft when this happened.[48] His recollection is supported by Samson who recorded that *Short was plugging away at seaplanes and the next one he made was a 70 h.p. Gnome Tractor with a single float. This was very difficult, in fact, dangerous to handle on the water in any wind.*[49]

The aircraft reverted to a wheeled undercarriage for the September 1912 Army Manoeuvres. From October it operated from the temporary seaplane station at Carlingnose as a single float seaplane. Cdr C.R. Samson took No. 5, Short Tractor No. 10 and Henry Farman Pusher No. 11 to carry out experiments on how aircraft could spot submarines. These were held in the Firth of Forth over three days. *No. 5 fitted with a single float was found useless in gusty weather or strong winds.* No. 5 capsized there on 4 October 1912, and was written off. *When No. 5 capsized, one piston of the 70 hp Gnome was found to have cracked, due to the sudden immersion in cold water.* Bad weather seriously hampered these experiments.[50] The S.41 was delivered after this machine.

The Short Brothers' Military Tractor Biplane Type S 45 was featured in the Shorts' catalogue. This described it as designed for military reconnaissance work. *The type is usually fitted with a "Gnome" engine, but other types of engine can be fitted if specified. The machine is a two-seater biplane, with a tractor screw, and a fuselage with non-lifting tail. A wheel control is provided for the warping and elevating of the planes, and the usual foot-bar for steering.*

The pilot seat is in front, facing a roomy dashboard fitted with all necessary instruments, switches, sight-feed lubricators, etc., and engine controls in convenient positions. The passenger's or observer's seat is behind the pilot, and is so arranged to give a clear view of the ground below, whilst at the same time being well shielded from the propeller draught. Convenient receptacles are provided for maps, instruments, etc., in both pilot and passenger seats, which are also fitted with comfortable air cushions. The planes and fuselage are covered with a strong, light fabric, which possesses a permanently taut and glossy surface, which is not affected by changes in temperature and is quite

Left: The object of much interest the S.45 (Serial T.5) at Portsmouth during the Naval Review in July 1912.

Right: Wherever the seaplanes went they drew large crowds.

Left: Bringing the S.45 onto the beach at Eastney on 5 July 1912. The machine has been lifted up by sailors holding the long plank under the single main float, and a rope is attached t the undercarriage and being pulled by sailors and many small boys. The seaplane had been flown from Sheerness by Lt Spencer Grey and Lt Sheppard. This photograph has also been captioned by J.M. Bruce as having been taken at Portsmouth on 8 July 1912, Note the two figures on the left with a mark above their heads. G.S. Trewin and Shepherd without hat.

Left: Close-up of S.45. Note how the undercarriage legs are wider than the single main float.

Left: Serial 424 was a landplane version. Note kingposts above upper wing bracing the wing extensions.

Specifications Short S.45							
Source	**1**	**2**		**2**		**2**	
Engine	70-hp Gnomê	70-hp Gnomê		80-hp Gnomê		160-hp Gnomê	
Span	42 ft 0 in	42 ft 0 in	12.9 m	45 ft 0 in	13.72 m	50 ft 0 in	15.25 m
Length	35 ft 6 in	35 ft 6 in	10.9 m	35 ft 6 in	10.9 m	40 ft 0 in	12.2 m
Height		11 ft 6 in	3.5 m	11 ft 6 in	3.5 m	11 ft 9 in	3.58 m
Wing Area	450 ft^2						
Empty/Net	1,080 lb	1,080 lb	490 kg	1,100 lb	500 kg	1,890 lb	860 kg
Loaded	1,500 lb						
Speed max	60 mph	60 mph	97 km	70 mph	113 km	74 mph	120 km
Speed min		50 mph	80 km	50mph	80 km	56 mph	90 km
Fuel Capacity		5		5		6	
Persons seated		2		2		4	

Source: 1. *Shorts Aircraft since 1900*. Barnes gives the same specifications as *Jane's All the World's Aircraft 1913*.
2. *Catalogue of Naval and Military Aeroplanes.* Short Brothers, Catalogue A. No date.

Left: The T.5 readying for take-off before a crowd. Lt S.D.A. Grey with Lt Philip Shepherd in rear cockpit. Margate, 30 or 31 May 1912.

Right: With the serial above the Shorts' logo, No. 424 shows the relationship to the earlier T.5. The lower wing is closer to the fuselage. The front panels are flat with a bulge to contain the 80-hp Gnôme rotary engine.

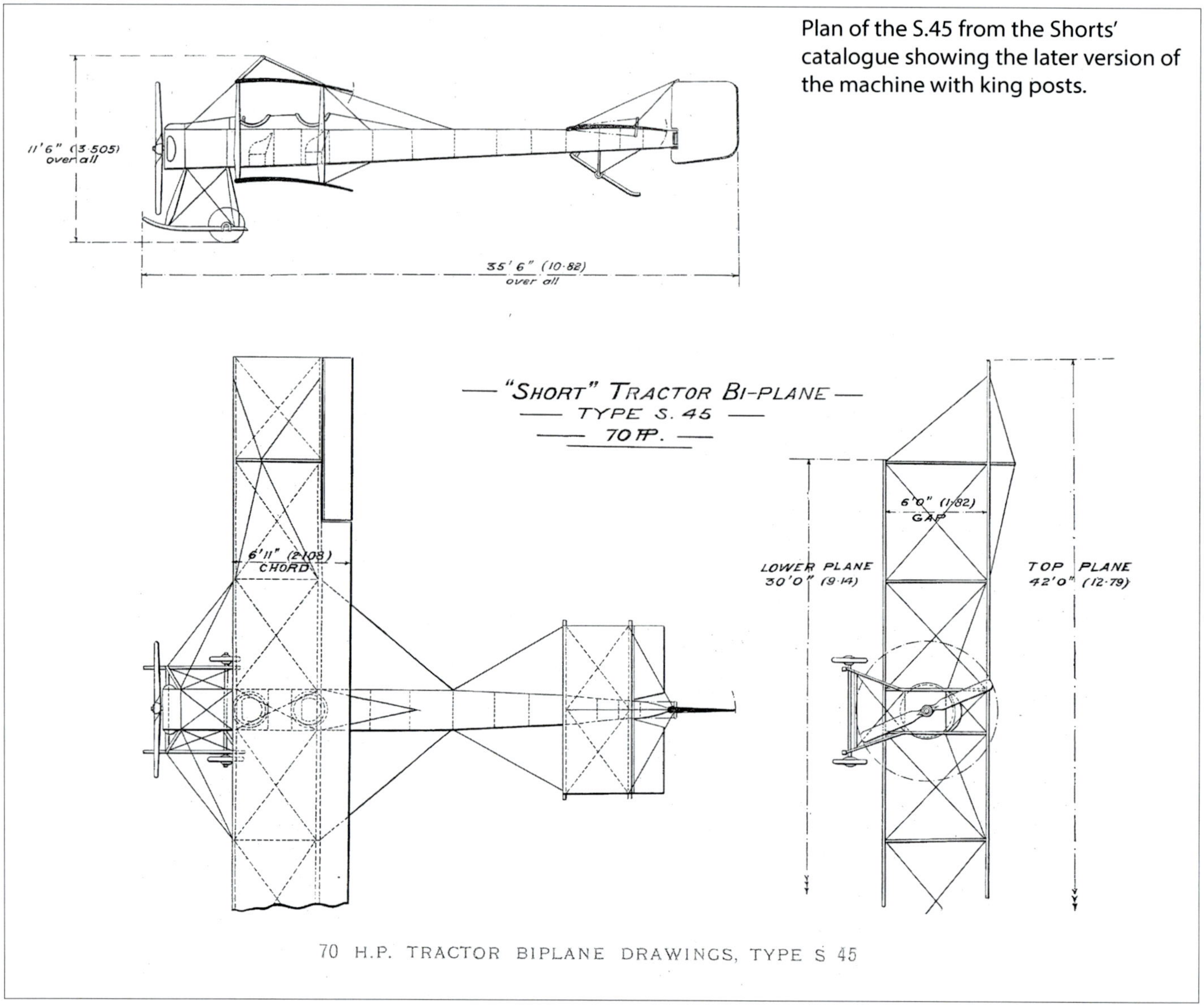

Plan of the S.45 from the Shorts' catalogue showing the later version of the machine with king posts.

impervious to moisture.

The landing chassis, without being unduly heavy, is of extremely strong and simple construction, being capable of withstanding rough landing grounds.

The utmost accessibility is provided to the engine, and, in fact, to all working parts which may have to be examined, this accessibility being one of the special features of all "Short" aeroplanes.

The material used are best English white ash and spruce for the woodwork, whilst all the fittings are of the best tough steel, either brazed or welded as most suitable for the strains the parts are called upon to bear.

This machine can be fitted with wireless telegraphy apparatus if desired, price upon application.

Jane's All the World's Aircraft 1913 describes the S.45 as a military tractor with tandem seats, the pilot in front, and fittings for maps, etc.

S.45 Type Tractor. RFC Nos. 413, 423, & 424

The War Office ordered three of the Short tractors as trainers for the CFS with serial Nos. 413, 423 and 425 (S.51, S.49 and S.50 respectively.) S.51 was delivered on 1 October 1912, to Farnborough. On 28 November, it achieved a new altitude record for the CFS, Lt R.R. Smith-Barry achieving 7,000 feet. No. 413 was written off on 3 December 1912, when Lt T O'Brien Hubbard, with Stoker Edwards as a pupil, came down at a steep angle and crashed.

S.49 and S.50 were delivered in December 1912. No. 424 was apparently written off when Maj E.L. Gerrard flew the biplane from Farnborough to Upavon on 17 February 1913. He had made a good landing in a gale when the machine was blown over. No. 423 flew to Upavon on 22 February again by Gerrand. It made some instructional flights on the 25th. Nothing further is known of these two trainers. However, the fact that the serials re-appeared on two B.E.8 biplanes is proof that the RFC had finished with the two machines.[51]

Specifications:
Span 42 ft; Length 35 ft 6 in; Height 11 ft 6 in; Wing area 450 ft^2; Weights Empty 1,080 lb; Loaded 1,500 lb; Speed 60 mph.[52]

Short S.47 'Triple-Tractor' Biplane. RNAS No. T4/4

Another example of the Short brother's desire for more horse-power, this landplane had two 50-hp Gnôme engines, one behind the other. They were mounted under a 16 foot long cowling, driving three tractor airscrews. One was in the usual position forward, and two were mounted on the first interplane struts rotated by chain drives. The two engines rotated in different directions. This was probably done in order to ensure that the air-cooled rotaries received enough air, however the heat generated by this installation earned it the nick name of the 'Field Kitchen.' McClean undertook the official test flights for the Admiralty on 24 July 1912, at Eastchurch. T4 became the mount of Lt Cecil John L'Estrange Malone, who flew it in the 1912 Army Manoeuvres.[53]

At Eastchurch it was used for trials with a Rouzet wireless installation. Lt Fitzmaurice recorded that he fitted the Rouzet transmitter to *one experimental twin-propeller machine, which no one liked; this machine was, however, the only possible one to which an apparatus could be fitted, and even then it had to be fixed to a platform outside the fuselage and streamlined in.* [54]

Lt Raymond Fitzmaurice wrote in January 1913, that *On the return of seaplane No. 4 from Dover I refitted the wireless apparatus and improved the balanced aerial capacity. I also tuned up the wave length to 700 feet, i.e. the Destroyers wave. A shittle belt was substituted for the belt supplied with the apparatus and extra insulation was given to various leads, &c. in the machine.*

Captain Risk R.M.L.I. (pilot) and myself proceeded up on her on the after noon of the 22 January to carry out trials. I made a signal addressed to any destroyer that might be within range to ask them to inform me if the signal was heard so that I could get some indication of the range of the apparatus using the Destroyer

Left: The pilot of the Triple Tractor must have had great difficulty is sighting over the hood that covered the twin 50-hp Gnôme engines. (via P Jarrett)

Right: The S.47 with the serial T4 on the rudder. Horace Short is standing in front of the wing. (via P Jarrett).

wavelength.

Fitzmaurice referred to a letter that he had received from the CO of HMS *Ouse* that showed that the signals were good a t 40 miles.

Damaged on 22 January 1913, it was apparently never repaired or reconstructed.

Specifications:
Span 48 ft; Length 41 ft; Area 500 ft^2; Speed 60 mph.

Short S.42 Tractor Monoplane. RNAS No. M2/8

McClean ordered two monoplanes but only one was completed and that for the Admiralty. It appears that McClean never flew the monoplane.

In their issue of 2 March 1912, *Flight* reported that Short Brothers, *who have hitherto only been associated with machines of the biplane class, have constructed a monoplane which was tested for the first time on Saturday morning last, by Commodore Samson. That is should fly satisfactorily at the first attempt and need no subsequent adjustments, is testimony enough of the remarkable excellence displayed by that firm in both design and construction.* Superficially this new machine, the M2, had much in common with the Bleriot.[55] The machine was featured in detail in the next issue of the magazine together with a two-view dimensioned plan of the monoplane.[56]

On 24 February 1912, the M2 was delivered to Eastchurch, and at 9 a.m. Commander Samson flew the *New Short Monoplane M.2. 10 mins 250 to try machine. Flew very nicely. Tail requires to be slightly less angle.* Later that day he flew it again to the middle of the harbour and back (11 mins). *Climbed beautifully.*[57] The following day he put the 50-hp Gnôme powered monoplane through its acceptance tests. Originally M2 it became serial No. 8.

Samson must have been pleased with the little monoplane as on the 27 February a Minute requested permission to conduct experiments and mentioned the B.3 (Short S.39 'Triple-Twin' Tractor/Pusher Biplane) and the M.2. *As soon as "B.3" and "M.2" are fitted with floats, and, when the tents, about to be ordered from Piggott Bros have been delivered (i.e. in about one month), I propose to erect one of the latter at Shotley for the temporary housing of the 2 mc/s which would be flown over from Eastchurch…*

It wd (sic) *then be possible to experiment with the Bomb Dropping Device now being elaborated by Cdr Samson, R.N., and Mr Horace Short.*[58]

Richard Bell Davies recollected a Shorts early effort as the 'little dirty.' *She was a single engined monoplane rather like a Bleriot, but something was wrong with the tail design with the result that if the engine was cut, her tail dropped. She would go up but would only come down either with the engine full on or tail first. Short wanted to take her back but we kept her for instruction on the anatomy of the aeroplane.* [59] This may be the reason for the entry in Samson's Log Book for 6 May 1912: - *8 AM 5 mins in M2 6' turned somersault landing*. Refurbished in September 1913, Samson recorded that he flew No. 8 in the evening of 13 March 1913, that the machine needed adjustment *flying very much right wing down (had hard job to land her) as tail would not come up.*[60] No. 8 was deleted before July 1914 due to its deterioration.

Specifications:
Span 29 ft 3 in; Length 25 ft. Wing Area 165 ft^2. Engine 50-hp Gnôme. Speed max 55 mph. Endurance 5 hrs.

Short S.46 Twin-engine Monoplane. RNAS No. H2/12

According to J Lankester Parker an incredible aeroplane was built for the Navy. This was *a tractor type. It was powered with two 50 h.p. Gnômes mounted in tandem. The front engine was directly coupled to the propeller in the normal way; the rear one was mounted in the fuselage. Gnôme rotary engines had a habit of throwing castor oil in all directions, which I suppose accounts for it being named the "dirty double".*[61] His recollection was correct except for the horse-power of the engines. W/Cdr S.T. Freeman recalled that it was *an enormous aircraft with one tractor & one pusher engine.* It was at Grain about 1912.[62]

This monoplane was ordered by the Admiralty as the H2, (later Serial No. 12), the H indicating a hydroplane.[63] It was powered by twin 70-hp Gnôme engines. No photographs or drawings are known to have survived, none were available when a presentation photograph album was prepared for Oswald in 1919, and Barnes acknowledges that the documentary descriptions are sketchy. The tail surfaces were carried on tail booms as per the usual pusher arrangement. It featured side by side seating and a short nacelle, such that the crew was doused with the castor-oil laced exhaust, this leading to the nick-name the *Dirty-Double.* Lt C.R. Samson made the first flight of the new monoplane at 0750 on Monday 23 September 1912, and recorded *10 mins on New Twin Engine Monoplane (H2)*, followed by another 5 minutes at 0810 that morning. On the 25th he recorded: - *6.0 PM. 10 mins on No 12 (twin Monoplane) had to land in marshes owing to faulty engine giving out – damage. Right Skid broken. Xxtype box crushed. Left axle bent. Very steady in moderate wind. 10'.*[64]

In October 1912, C.G. Grey of *The Aeroplane* was at Eastchurch and was able to visit the Shorts' works where the *triple tractor and triple-twin Short biplanes were being overhauled, and the great 140-h.p. experimental monoplane with tandem engines, was having some repairs and alterations made, so a certain amount of interest was lacking.*[65]

On 23 October 1912, Samson recorded in his log that the *Twin Monoplane* skid and propeller were damaged in *landing on water.* Edward Featherstone Briggs recorded in his log that he was flying with Samson on 24 October when they landed on *big floats* breaking the propeller and front struts needing the machine to be towed to Grain, meaning it

This Page: The Short monoplane M2. Its resemblance to the Bleriot XI may be readily seen in these views. (via P Jarrett)

was buoyant. Notwithstanding the dates being different, this would be the same incident.[66]

Briggs noted that the machine was *Very steady, rather like a floating raft* when flying as a passenger with Samson on 23 October 1912, and when he flew the *Twin mono for the first time … Quite easy to fly and land keeping front engine going.*[67] This is different to McClean's view of the S.46 as a *very bad machine.*[68] To date there has been no documentary evidence that the S.46 ever took off from water. It was at Grain before being returned to Eastchurch around early 1913, and was dismantled during the W/E 2 August 1913.[69]

Short S.39 'Triple-Twin' Tractor/Pusher Biplane. RNAS T.3/No. 3

McClean ordered a twin-engine biplane before he left for the Pacific on the solar eclipse expedition of 1911. James Lindsay *Jim* Travers joined Short Brothers as Horace's assistant and designer. According to Travers, Horace had been working on the design of the twin-engine machine for months and had little success. Drawings had been shown at the 1910 Olympia Aero Show and had received favourable comment in the aviation press, however, whether the machine that finally emerged resembled the plans is unknown. Travers wrote that Horace was continually making changes to the S.39.[70]

The first twin-engine and the first 'in-house' design by Shorts, the S.39 was a milestone for the Company. The S.39 was structurally the same as the Farman-Sommer types built by the company with a nacelle containing the two engines, one 50-hp Gnôme in an overhung pusher mounting at the back and another 50-hp Gnôme at the front driving two airscrews mounted outboard and driven by chains. The airscrews rotated in opposite directions. The cockpit was set up for side-by-side seating.

The wings had single acting ailerons on both top and bottom wings. Extensions were fitted later and these also included the ailerons. In October 1912, the extensions were removed and wings of 50 feet span fitted. Later more extensions were added to give a span of 64 feet.

First flown on 18 September 1911, by its owner, Frank McClean, it made a brief flight then McClean took Samson up for a flight around Eastchurch aerodrome. McClean throttled back each engine in turn, experiencing for the first time the effect of having ample power. The S.39 was flown by Naval trainees at Eastchurch. By mid-June 1912, it bore the serial T3 after purchase by the Admiralty in June 1912. On 24 July 1912, at Eastchurch, the *New Naval 100-h.p. triple tractor Short biplane flew the one hour test with Mr. Frank McClean as pilot and Lieut. Malone as passenger, successfully accomplishing the test.*[71] The machine was later entirely rebuilt as S.78, a single engine, tandem pusher biplane. (See above under the S.38 types).

Specifications:
Span 34 ft (later 50 ft); Length 45 ft; Area 435 ft^2 (later 500 ft^2); Weight Empty 1,800 lb; Loaded 2,100 lb. Speed 55 mph.

Short S.40 Biplane

The S.33 biplane was built for Frank McLean and, from research since Barnes published his *Shorts Aircraft since 1900,* it has been established that the biplane that was flown by Cecil Grace when he disappeared over the Channel was not the S.40 but S.33.

The aircraft that McClean made his famous flight up the Thames on 10 August 1912, was the S.40.[72] The S.40 had been converted as a seaplane with twin-floats. The floats were flat bottomed pontoons without a step. Construction was a wooden frame with waterproof canvas covering and braced

Left & Above: Shorts' twin engined S.39 with extended upper wings. The landing gear is quite complex. No knowledge seems to have survived on the operation of the fuel system. But with four tanks on the interplane struts it musts have been complicated for its day.

internally with piano wire. McClean made the first take-off and alighting by a Short seaplane in this aircraft sometime between 30 May and 9 August.[73]

McClean flew under all the bridges just above the water and through the opening of Tower Bridge between the footbridge and the road span, touching down on the water between Charing Cross and Westminster Bridges. On the following day the police prevented McLean from repeating the flight in reverse and insist he taxi down to Shadwell Basin for his take off. The S.40 was damaged in trying to take off from there and returned to Eastchurch by road.

The following from the issue of *Flight* for 14 August, describes McClean's flight and subsequent accident.

By Hydro-aeroplane up the Thames

ALTHOUGH London was deprived by the appalling weather of the sight of M. Beaumont piloting his hydro-aeroplane up the Thames, the visit of Mr. F. K. McClean more than compensated for the loss. Remembering an appointment in town on Saturday -morning, Mr. McClean thought it would be a good idea to come up on his Short machine, and so at 6 a.m. he had it brought out of its shed at Harty Ferry, in the Isle of Sheppey, and after seeing everything in order he started off. Following the coast round Leysdown, Warden Point to Sheerness, he continued over the Thames. At Gravesend the smoke of the various factories rather troubled the aviator but he made good progress. Approaching London Mr. McClean brought his machine lower down and negotiated the Tower Bridge between the lower and upper spans, but the remaining bridges to Westminster he flew underneath, the water being just touched at Blackfriars and Waterloo bridges. He reached Westminster about 8.30 and was taken ashore to Westminster Pier on a Port of London Launch. The return journey on Sunday afternoon was not so successful - owing to restrictions as to rising from the water which had been imposed by the police. The bridges had all been safely negotiated, and when near Shadwell Basin Mr. McClean started to manoeuvre to get into the air at the point designated by the river authorities. He had made one circuit when the machine side-slipped, and either through hitting a barge or by sudden contact with the water one of the floats was damaged. The machine was then towed into Shadwell Dock, this operation being superintended by Mr. McClean from the driving seat, and dismantled for its return by road to Eastchurch.[74]

The S.40 had been used by the Territorials before McClean's famous flight.

It was later reconstructed as the S.58 and was completed by 15 November 1912.

Short S 41 Tractor Biplane Hydro Aeroplane RNAS No. H1/10

See Chapter 3.

Short S.48 Hydro-monoplane with twin tandem 70-hp Gnôme engines

Ordered by McClean this machine was cancelled and not built.[75]

Short S.52 Twin-Engine Tractor Hydro Monoplane. RNAS H3/14

This proposed monoplane powered by two 140-hp Gnôme rotaries was not built.

Short S.53 Twin Engine Tractor Biplane

Another of McClean's orders that was not built, no evidence having been found of this machine. The biplane was to have two 70-hp Gnôme engines.

Short S.54 Tractor Seaplane. RNAS No. 19

Ordered in August 1912, S.54 received the Serial No. 19. It was ordered with a 140-hp Gnôme, but by February 1914, it had a 160-hp Gnôme installed. It is not known if it ever received the 140-hp engine. It was a spare machine for the Royal Review in July. Damaged in a take-off on 14 August, it was sent back to Shorts for repair and was recorded by Grain as arriving in tow from Queensboro after being rebuilt by contractors on 6 November 1914.

Short No. 19 from the Isle of Grain under Flt H.G.

Right: Short S.54 Serial No. 19 on a sandy shore on its movement trolley.

Left: Poor photograph of No. 19 on the shore where there is the usual gathering of people when these wonderful marine aircraft came ashore.

Right: The only markings carried by RNAS aircraft at this stage was usually the serial on the rudder. No. 19 also carries it under both wings. National markings have yet to be applied, however the war would soon make the markings of friend or foe necessary.

Wanklyn and L/Tel Leading Mechanic R.L. Hartley made an emergency alighting on the North Sea on 14 December 1914, and was found 3 miles NE of the Kentish Knock by the Netherlands ferry SS *Oranje Nassau* from the Zeeland company, on its way to Vlissingen. The *Oranje Nassau* picked up the crew and returned them to England on 15 December.

The aircraft was found by the trawler *Ocean Rambler* and it was transported to Harwich on the deck of HMS *Brazen*. The engine was salvaged, the aircraft being written off on 22 December.[76]

Short S.56 & S.57 Tractor Biplane Hydro Aeroplane RNAS Nos. 20 & 21

See Chapter 3.

Short S.60 Olympia Aero Show Tractor Biplane Seaplane. RNAS No. 42

Messrs Short Brothers will exhibit on Mr. Percy Grace's stand a tractor hydro-biplane, resembling in general design the firm's famous S41 type, which was flown with such success during the Naval Review at Weymouth in May, 1912. The machine is supported by two main floats, arranged catamaran fashion, and a third float beneath the tail. Small balancing floats are placed at the wing-tips also. The engine is an 80-hp Gnome, which can be started up from the passenger's seat, a feature of manifest value in a hydroplane. Seating accommodation is provided for three persons, the pilot's seat being placed behind the two passenger's seats. The main dimensions of the machine are as follows: span 48 ft top, 30 ft bottom.; chord, 5 feet; gap, 5 feet; length (over-all), 34½ feet; height, 11 ½ feet. *The biplane tail is provided with ailerons for lateral balancing; the tail is non-lifting. Messrs. Short Brothers have devoted a great deal of attention to the development of this type of hydro-aeroplane, and all the trials of this promising production will be watched with considerable interest.*[77] This machine was the S.60.

The Short S.60 was exhibited at the 1913 Olympia Aero Show. *Flight* reported that steel tubular stream-line interplane struts were fitted in place of wooden struts. The fuselage was a simple girder of rectangular section. *In front under an aluminium cowl, rotates the 80-h.p. Gnome motor that drives a large diameter Integral propeller. The engine is carefully shielded in, both on top and below, so that it may not become splashed by any spray that may be thrown up by the floats passing through rough water.* The passenger's seat could be moved to right to allow for another passenger. Wheel control was provided for the pilot who was seated behind the passengers. The main fuel tank was on the floor at the approximate centre of pressure of the machine, from there it supplied a service tank under the motor cowl by a small wind driven pump. Endurance was 6 hours.

The front spar was an ash H-section, whereas the rear spar was rectangular ash. Spruce ribs were used. The bottom wing was now mounted directly on the fuselage to give greater structural rigidity. The interplane struts were steel tubes. The pilot occupied the rear seat. Two passengers could be accommodated in the front seat, side by side, and provision for a starting handle was provided for the passenger.

The floats were simply plain pontoons made of wooden frames covered by heavy fabric, and were supported by stout steel struts. The wingtip floats were made of cylindrical form from canvas shaped around an internal steel strip such that if punctured they will still retain their shape. A float with water rudder was fitted to the tail. The water rudder acted in conjunction with the rudder. The tail was non-lifting of conventional construction. The machine had a maximum speed of 65 mph.[78]

Purchased by the Admiralty the S.60 received the serial No. 42. On Sunday, 27 July 1913, Maj Gordon, RMLI, with Lt Babington, RN, *alighted rather heavily, bursting two floats, and were towed in by torpedo boat 36. As a result no flying was done last week at Leven until Friday, when all four officers had a trial flight.* Gordon and Babington flew to the other side of the Forth and alighted at Ketsroe. Returning on Saturday morning, the chain of the starting gear broke and the machine was again towed into shore.[79] The floats were temporarily replaced by a land undercarriage in January 1914, and were refitted when No. 42 took place in the 1914 Naval manoeuvres.

The Aeroplane reported that on Tuesday, 20 January 1914, No. 42 alighted off St Andrews under Maj Gordon, RMLI, with Capt Barnby, RMLI. In attempting to get off the floats were so damage that the machine had to be taken ashore where it was converted into a land machine. Maj Gordon *had some trouble with the controls while putting the machine through its tests over the sand, but it got away safely.* They arrived at the Dysart Aerodrome on Wednesday and the following day, flew to the new aerodrome at Broomfield, where the machine was housed[80]. By early March, No. 42 was flying from the naval base at Carolina Port, Dundee, which was *temporarily a land base, as all the flying has been done on Short No. 42 with a land chassis. The sheds are still being put up, but it will be some time before they are ready. They are placed about 100 yards from the banks of the Tay, so that a long slipway will have to be constructed.*

A report for Thursday, 19 March 1914, noted that the Short No. 42 arrived from Dundee, by air, at Leven. Two Short seaplanes had arrived in boxes the previous week and the Borel seaplane No. 86 was expected. The Short seaplanes were probably Nos. 74 and 77 (100-hp Gnômes) that were being erected, and a third started (probably No. 77), during the W/E 26 March. All hands were engaged in overhauling and setting everything in order for the manoeuvres that lasted till 3 April. Although the weather was unfavourable, practically all the machines flew on service. All the Short seaplanes, Nos. 42, 74, 75 and 77 had been fitted with wireless. Their work was mainly reconnaissance flying from the mouth of the Firth to Rosyth. It was reported on 30 April that all the machines had left Leven for Dundee and the Bessonneau hangars were dismantled and only the barbed wire fence remained to show that there had been a camp at the site.[81]

On 22 August 1914, Nos. 74 and 77 were being prepared at Sheerness with spare engine and all necessary stores for transhipment to the SS *Indrani*. On the arrival of the ship, it was discovered that in order to stow the machine onboard with any safety, Nos. 74 and 77 would have to be dismantled. The carpenters worked all night dismantling the two machines and preparing them for shipment. The ship left the next day with the aircraft and spares for Yarmouth Air Station. The same day the HMS *Empress* arrived at Sheerness from Ostende with Nos 75 and 76 on board, apparently for repair.[82]

On the outbreak of war, Samson wrote that after *a period spent on the East Coast of England carrying out patrols against possible air attacks, my unit of the R.N.A.S. was ordered to proceed overseas, to co-operate with a Brigade of Royal Marines who were to occupy Ostend.* On 27 August 1914, the unit went to Ostend. *As for the aeroplanes, the majority were old veteran servants of the Crown and I doubt that anybody has ever taken to war such a heterogeneous crowd of machines. … they consisted of:*

- *2 B.E. Biplanes, fitted with a 70-h.p. Renault engine.*
- *2 Sopwith Biplanes, fitted with an 80-h.p. Gnome engine.*
- *1 Blériot Monoplane, fitted with an 80-h.p. Le Rhône engine.*
- *1 Bleriot Monoplane, fitted with a 50-h.p. Gnome engine.*
- *1 Henry Farman Biplane, fitted with an 80-h.p. engine.*
- *1 Bristol Biplane, fitted with an 80-h.p. Gnome engine.*
- *1 Short Biplane, fitted with an 80-h.p. Gnome engine.*

The last in this list, the Short biplane, was originally a seaplane, but its floats had been removed and a wheel chassis put in their place.[83]

Samson moved his HQ to Morbecque. He intended to keep the HQ flight there with the intention of carrying out patrols over Douai, Cambrai and Valenciennes. *This sounds grand, but at this time only consisted of No. 42.*[84] Short No. 42 did not last long. Samson takes up the story again: -

On the 28th (of September) *I had a crash in old No. 42. This aeroplane was really on her last legs; but as we were so woefully short of aeroplanes we had to use her. As soon as I left the ground the engine stopped dead. I was faced with the alternative of attempting a turn which would have landed her in some trenches full of Marines or of charging the trees in front of me.*

I chose the trees, and luckily managed to find a soft one. The aeroplane remained hung up in the branches about 30 feet in the air, and I escape without a scratch; but No. 42 had flown her last flight.[85] She was officially deleted on 14 November 1914. Although she was unique, the S.60 may be regarded as the type linking the S.41 and the 'Improved Type S.41', the Admiralty 74 Type.

Specifications:
Span Upper 48 ft; Span Lower 30 ft. Length 35 ft. Wing Area 390 ft^2. Weight Empty 1,200 lb. Weight loaded 1,970 lb. Speed max 65 mph. Engine 80-hp Gnôme.

Short S.68 Circuit of Britain Contest Tractor Biplane Seaplane. RNAS No. 182?

The *Daily Mail* offered a prize of £5,000 for the first aircraft to complete a prescribed circuit round Great Britain in flight within a period not exceeding 72 hours. Pilots had to be British and hold an aviator's certificate issued by the Royal Aero Club. A passenger had to be carried throughout the flights. *The complete aircraft and all its parts, including the motor, must have been entirely constructed within the confines of the British Empire, but this provision shall not be held to apply to raw material.*[86]

McClean ordered a machine to enter in the Circuit race. His machine *for the "Daily Mail" Waterplane Circuit is now well under way at Messrs. Shorts factory. The Advent of this machine will be watched with interest, as the water planes hitherto turned out by Short Bros. have been so successful. It is to be fitted with a 100-hp Green engine, and is expected to develop a speed of sixty miles an hour,* reported *The Aeroplane* of 31 July.[87] Unfortunately, the British engine was to prove troublesome.

The S.68 with its racing number '4' on the rudder, was modified with lower wing extensions and both wings increased in chord by extending their leading edges. However, the machine was not successful and McClean withdrew. In a 1914 list of possible aircraft that could be impressed by the military, the S.68 was recorded. It was to have a new engine fitted. Jarrett suggests that this may have been the Short 'Dual Control Seaplane' with 100-hp Gnôme ordered under Contract CP44634/14 as No. 182, but never delivered.[88]

Right: Serial No. 42 Short S.60 at Leven, 16 July 1913. Beyond the roped area there is the usual crowd of onlookers. No. 42 was delivered to the temporary station at Leven on 12th July.

Left: Serial No. 42 Short S.60 at Leven, 16 July 1913. Although captioned the same, in this photograph the machine now has serials applied under both lower wings.

Right: No. 42 with land undercarriage. Montrose, 1914. Originally it was a temporary lash-up. Back to floatplane for 1914 Fleet manoeuvres. Reverted to landplane. After the war commenced went to Belgium with what was to become No. 3 Squadron, RNAS. Cdr C.R. Samson hit a tree in this machine on 28 September 1914, Morbecque. Damaged beyond repair.

Left: Serial No. 42 on the beach.

Specifications:
Span 61 ft; Length 40 ft; Area 600 ft^2; (later 660 ft^2).

Short S.69 to S.72 Admiralty 74 Type Tractor Biplane Seaplanes. RNAS Nos. 74 to 77

See Chapter 4.

Short S.73 & S.74 Tractor Biplane Seaplanes. RNAS Nos. 78 & 79

A.H. Sandwell was appointed to Dundee after completing his training at Eastchurch. *The seaplanes with which the Dundee base was equipped when I got there in July 1915, could be numbered on one hand – but it took two hands, a pair of feet and a faith that could move mountains to fly them. Two Sopwiths, with the then new and more or less untried 225hp Sunbeam engines, two Avros with 150hp engines of the same make, and an early Short – No. 78 – with a 160hp (twin 80 Gnome rotary) completed the establishment.*

Sandwell found that the *Gnome-Short was so pleasant, (and so fragile), that it was more or less reserved for a few of the senior officers.*[89]

The seaplane Sandwell was referring to was one of two built to Contract CP38479/13. They were basically similar to the Type 74 but had the twin-row 160-hp Gnôme with exhaust funnel. No. 78 is thought to be the machine that Lt J.W. Seddon flew Winston Churchill from Tilbury Fort to the Isle of Grain on 7 February 1914. After the outbreak of the war, No. 78 was recorded by Grain as patrolling to Ostende on 9 August 1914. No. 79 was 'in readiness.' No. 78 was ordered to Dundee on 12 September 1914, and reported as 'Ready' on 17 January 1916. It did not last much longer, being deleted on 9 May 1916.

No. 79 performed a search for submarines on 10 August 1914, from Grain. It proceeded on a cross Channel patrol and ended at Dunkirk. No. 79 was also at Dundee, leaving Grain for that station on 21 September 1914. It was lost on 1 January 1915, when it was forced to alight due to engine trouble. The crew of Lts H.A. Busk and L.H. Strain were rescued. The wreckage was washed ashore at Sanmuir on the 2nd. Officially deleted on 17 January.

Left: The Short S.68, Frank McClean's entry for the 1913 Daily Mail Circuit of Britain endurance race. The large vertical radiators are noteworthy. Although it is shown with race number '4'. McClean withdrew as the 100-hp Green engine did not give an adequate performance.

Right: The S.68 had the wings modified to try to improve its performance. The lower wing was increased in span to match the upper wing and the leading edges were extended on both upper and lower wings.

Short S.79 Two-seat Pusher Seaplane. RNAS No. 80

Built under Contract C.P.38479/13, it received the Serial No. 80. Power was provided by a 100-hp Gnôme-Monosoupape rotary engine. Delivered to Grain in January 1914, it is thought to have been stored until it was re-erected in December 1914. The seaplane survived until September 1915.

Lt William Geoffrey Moore came across No. 80 when he was sent to the Isle of Grain. Although a graduated service pilot, Moore had never been taught to fly a seaplane. As an active service station, Grain had no one to teach him to fly a seaplane and so he was given use of *a venerable old lady,* Short No. 80. Underpowered with a 100-hp Gnôme engine, the machine would only come unstuck in a flat calm. Moore had to teach himself to fly a seaplane by himself in No. 80.

He was usually accompanied by Lt Henry M Cave-Brown-Cave, RN[90], who, as the station Engineering Officer, was not supposed to fly, being considered too valuable to risk flying. The two would sit side by side in the wide nacelle and had a hilarious time bumping around in No. 80.[91]

Short S.80 Nile Pusher Side-by-side Biplane Seaplane. RNAS No. 905

In September 1913, McClean ordered the S.80. A twin boom pusher with a 160-hp twin-row Gnôme, side by side seating and the 67 ft span wings folded. For the story of the Nile flight see Phillip Jarrett's wonderfully illustrated *Frank McClean, Godfather to British Naval Aviation.*

In July *Flight* reported that McClean *Made a fine flight on his Short hydro-biplane, 160 h.p., at Harty with a passenger; this was the first time the machine had been out since its return from Egypt.*[92] Apparently, McClean offered the seaplane to

Right: Short No.78 under armed guard. The addition of the Union flag marking would indicate that the war had commenced before this photograph was taken. The aircraft does not appear to be damaged.

Left: No. 78 at Westgate seaplane station, August 1914. Holiday makers flocked to St Mildred's Bay to see the floatplane after it was pulled up from the beach to the grassy slope on the clifftop. Just forward of the aircraft the *Astra Torries* airship may be made out.

Right: With a small serial '78' and late markings, this is probably No. 78 after its engine was replaced with a 100-hp Gnôme.

the Admiralty on the outbreak of the war, but they impressed it instead and paid him £1,200 for it under Contract CP65176/14. McClean flew the S.80 to Grain on 2 August and left it for the Navy's use.[93] First used at Grain as a dual control trainer, it was allocated to HMS *Rivera* after being fitted to carry 100-lb bombs and a torpedo. Reallocated to *Hermes* in September 1914, it flew out to the carrier on the 28th for hoisting in trials, but the strong gusty wind made it inadvisable to try and the machine returned to Grain. Its ultimate fate is unknown; however, it may have been sunk when *Hermes* was torpedoed by U-27 on 31 October, or blown up when the aircraft's bombs exploded prematurely a few days earlier. At the time *Hermes* was being used to transport seaplanes to Dunkirk. Deleted in November 1914.

Short S.81 Gun-Carrying Pusher Biplane Seaplane. RNAS No. 126

The Committee of Imperial Defence, Royal Flying Corps, 1st Annual Report of the Air Committee, of 7 June 1913, under

Above Left: Short No. 79 in flight showing the chimney that the 160-hp Gnôme required.

Left: The non-folding Short No 79 (S.74) is pushed into the sea.

Left: No. 79 on the beach front. This machine made cross Channel patrols after the outbreak of hostilities. It was wrecked in a forced alighting in heavy seas on 1 January 1915.

Right: The S.80 on the Nile at Cairo where it was assembled. Horace Short is standing on the float. 3/4 January 1914.

Left: The S.80 beached for engine repairs while on Frank McClean's Nile River flight in 1914.

Right: The S.80 taxying.

Left: W.G. Moore's photograph of the S.80 as dual-control trainer Serial No. 905, at Grain in 1915. RNAS red ring insignia on the wings, but no serial on the rudder.

1914	
25 May	*Short machine No. 126 arrives.*
2 June	*No. 126 Short went for acceptance trials and passed tests.*
4 June	*No. 126 … went for practice flights carrying two passengers and additional weights. Found machine quite satisfactory.*
9 June	*Mr (*Horace*) Short visited station to inspect seaplane 126 which had damaged the landing gear on Thursday.*
25 June	Tuning engine of 126.
13 July	*Tested engine of No. 126, found full of oil.*
18 July	Returning from Naval Review *No.126 broke her propeller in the air.*
29 July	Tested with new propeller.
30 July	Departed Calshot for armament trials at Isle of Grain but broke propeller and petrol pipe at Ramsgate.
Source: Notes from A.M. Longmore's logbook by Gordon Bruce. (Logbook in National Maritime Museum, Greenwich, London.)	

the sub-title 'Guns for use in Aircraft,' reported that as *far as guns for use in aircraft are concerned, there is a 1-pr. Q.F. gun manufactured by the Coventry Ordnance Works now undergoing trial by the Ordnance Board.*

A 1-pr. Gun manufactured by the Elswick Ordnance Company is overdue for delivery, as is also a 1½-pr. Q.F. being manufactured by Vickers, though delivery in the latter case has been delayed by the embodiment of improvements suggested by the Ordnance Board after inspection of a mock-up model.[94]

The Short S.81 Gun-carrying pusher biplane was ordered under Contract No. CP.01764/13. The RFC (Military and Naval Wings) had ordered the three heavy semi-automatic guns for trials and the S.81 was specially designed to carry the Vickers 1½-pounder.[95]

The S.81 received the serial No. 126 and was a large three-bay pusher biplane powered by a 160-hp Gnôme engine with four-bladed propeller. The twin rudders and horizontal tailplane were carried on outrigger booms. The nacelle for two crew members was specially designed by Arthur Camden Pratt under Horace Shorts instructions. The wing extensions were braced by struts running from the lower wing to the upper wing. The S.81 introduced rubber-sprung floats, an innovation of Oswald Short. Wingtip floats were fitted due to the tendency of the machine to roll while moored out or while taxying.

Arthur Longmore recorded that Monday 25 May 1914, was fine with a 20-25 kt wind. There was no flying at Calshot that day but No. 126 arrived and was placed in the large shed. On 2 June, Gordon Bell put No. 126 through its acceptance trials and it passed. 2,000 feet in 21 minutes. On the 4th Lt Bigsworth took No. 126 up with two passengers and additional weight, for practice flights.[96]

Lt Arthur M Longmore, RN, was appointed to command the RNAS experimental station at Calshot on 13 January 1911. Longmore's logbook enables a time line for No. 126's armament experiments to be determined.

Various experiments were carried out in May 1914 with a machine gun. In October the armament was altered and the machine was tested in flight on the 17th to check the distribution of weights. No. 126 was fitted with a Lewis gun carrying seven magazines (drums) and also carrying a Maxim machine gun. The Lewis gun was tested in flight by Babington and Flt Lt Cave on the 18th.

A destroyer (H.M.S. Nubian) had been detailed for the purpose of a series of trials, and equipped with, and her crew trained in the use of, kites.

On the flying wire of a team of kites, a large flag was flown having been painted on it, a life size representation of an aeroplane pilot, and the engine and petrol tank of an aeroplane.

By steaming head to wind at full speed in a fresh breeze, this target could be made to move through the air at 60 miles an hour or more.

A seaplane machine built for a 1½ pounder gun, but carrying for these trials a maxim (sic) *gun only, flew past and attacked this target, scoring a number of hits on the depicted pilot, engine and tank, and many more on the flag.*

An extemporised telescopic sight, consisting of a naval gun sighting telescope, was used for this as well as for many of the previous trials, and the system of sighting by means of graticulations in the field of the telescope, on the principal used later in the Aldis gun sight, was in process of evolution, with the assistance of a private firm of opticians.[97]

A report on the test firing of the 1½-pdr Vickers gun that took place with No. 126 on 11 December 1914, (crew of Flt Cdrs Babington and Hyde Thompson), noted that the gun was fired at targets in the air and on the sea. Two shots were fired in the air and two on the water with very little shock to the structure or the pilot.[98] The seaplane was also fitted with a Lewis gun. The Vickers 1½-pdr was removed for this; as noted above, the machine was modified to be able to alternate the armament carried.[99]

Arthur Camden Pratt recalled that the S.81 was designed for a recoil of half a ton in several directions. He had left Shorts before the weapon had been tried, however he picked up a yarn afterwards that the S.81 stopped dead in the air when the gun was fired and dropped 500 feet before control was regained![100]

Lt R.H. Clarke-Hall was a naval gunnery officer and had also learnt to fly. In March 1913, he was appointed for duty with the RNAS and from that point the Admiralty and Ordnance Board pushed on with the subject of armament for aircraft with considerable vigour. C.G. Grey wrote that Clark-Hall was the first aviator to fire a cannon from an aeroplane. *Short Bros. built a big float biplane with a Samson (Canton Unné) pusher motor of 200 h.p. A 1½ -pounder Naval quick-firer was*

Right: The S.81 with its Serial No. 126 proudly carried on the twin rudders. The twin tail floats are well shown in this view. The armament is probably the 1½-pdr Vickers gun.

Left: Close-up of the S.81 showing the gunner's position and the pilot with his control wheel. The full wording of the stencilling on the front of the nacelle has not been determined.

Below: S.81 taxying with 1½-pdr Vickers gun.

Left: The S.81 at Great Yarmouth, has cockades added to the wings and a Davis recoilless gun mounted on the nacelle.

Below: The beached S.81, with a single rudder, awaiting recovery from Kessingland, 27/28 January 1915. The seaplane appears to be without a propeller. On 26 January 1915, Serial No. 126 was flying to Yarmouth but came down at Felixstowe due to a shortage of fuel. Continuing the flight, Lt C.F. Lan Davis came down off Kessingland and drifted ashore, the seaplane being badly damaged in hauling it up the beach. The seaplane was repaired at Yarmouth and continued experimental work until it was deleted on 20 October 1915.

Experimental Weapons (for Armament of Aircraft) 1914

1½-pr Q.F. Vickers	*Gun constructed for use in aircraft in accordance with conditions laid down by the Ordnance Board.*	*At Calshot for trial in seaplane 93 by RFC (Naval Wing)*
1-pr Q.F., Elswick Ordnance Co.		*At Calshot for trials in seaplane 61 by RFC (Naval Wing)*
1-pr Q.F. Coventry Ordnance Works		*Has been fired from an aeroplane slung from a gantry. No appreciable effect from recoil. Trials to be carried out at Farnborough by RFC (Military Wing), when aeroplane available. F.E.2 in which the trial was to be carried out, has been wrecked.*
Light Maxim, Vickers, Naval.	For trials in Naval aeroplane No. 66	Trials in progress. Several trials have been carried out with promising results.

Source: Committee of Imperial Defence, Royal Flying Corps, 2nd Annual Report of the Air Committee, 9 May 1914. P.30. TNA AIR1//362/15/231/3. (Note this report contains details of other armaments including bombs.)

C.F. Snowdon Gamble described all three of these QF guns as 'semi-automatic,' and that the semi-automatic gun was not suitable for the work but that an automatic gun was. (*The Air Weapon, Being an account of the Growth of British Military Aeronautics from the beginnings in the Year 1783 until the end of the Year 1929*. Oxford University Press, UK, 1931.)

Italian Offer

In 1914 an offer of the Short Folder seaplanes was made to the Italian Navy. *The Regia Marina, Direzione delle Costruzioni Navali, R. Arsenale*, Venice, noted the offer in June 1914, and was of the opinion that the structure was not different to those machines already purchased for the Navy and similar to the Breguet and Albatross biplanes.

The Navy's aviation school in Venice noted that except for the advantage of folding the wings, this aircraft does not present significant advantages over similar types already owned by the School (Albatros & Breguet).

Of the three proposed, the first of 100 hp Gnome 9 cylinders is not to be taken into consideration because given the conditions to which the device must satisfy, I consider the strength of 100 hp insufficient

On the other hand, given the good results recently obtained with fixed engines, I believe that rotary engines should gradually be abandoned.

On the other hand, the 135-horsepower and 200-horsepower Canton-Unné types could be considered if not that given their respective prices of 75,000, and 82,000 and 500, I believe that given the scarcity of funds allocated to aviation it is not for now to think to their acquisition.

Short seaplanes were not to be utilized by the Italian Navy in World War I.

Notes

1. Regia Marina, Direzione delle Costruzioni Navali, R. Arsenale Venezia. Signed by Il Colonnello del Genio Navale, Direttors. 27 June 1914. Copy via R Gentilli.
2. R Scuola di Aviazione Marina, Venezia. 15 July 1914. Copy via R Gentilli.

mounted in the nose and Clarke-Hall used to sit up behind it and poop off at targets in Southampton Water, or the Solent.

He was a very brave man because the kick of that gun was enough to shake the machine to pieces in the air. I never saw the thing fired myself, but my friends at Calshot used to say that every time he fired it the machine seemed to stand still, and then take a dive to pick up its flying-speed again.[101]

On 27 January 1915, No. 126 was badly damaged. Alighted at Kessingland near Yarmouth and drifted ashore.

On strength at Great Yarmouth during 1915, No. 126 was chosen for trials with the 6-pounder Davis non-recoil gun[102], and these took place on 13 to 16 April 1915. C.F. Snowden Gamble noted that the Davis gun weighted only 180 lb and *was an ingenious weapon where the shock of its recoil was absorbed by the discharge, in the opposite direction, a mass of small shot equal in weight to the projectile. The gun was double-ended, the breech-block being in the middle of the barrel. The trials, in so far as they went, were successful, but although large numbers of this gun were manufactured, it never came into general use in the Air Services.*[103]

A single rudder was fitted by January 1915. No. 126 was deleted on 20 October 1915.

Specifications:

Span: 67 ft; Length: 33 ft 9 in; Area: 540 ft^2; Empty Weight: 2,200 lb; Loaded Weight: 3,600 lb. Speed: 60 mph.

Short 74 Type Seaplane

See Chapter 4.

Short Admiralty Type 81 Three-bay Folder Tractor Biplane Seaplane. RNAS Nos. 89 & 90

These proposed seaplanes were to be powered by 160-hp Gnôme rotary engines. Cancelled by January 1914.

Short S.88 & S.87 Admiralty 135 Type Folder Tractor Biplane Seaplanes. RNAS Nos. 135 & 136

See Chapter 6.

Short S.90 – Type C (166 Type) RNAS Nos. 161 to 166

This development of the Short Folder type was intended for torpedo dropping. See Chapter 7.

Short Single-Seater Biplane. RNAS No. 145

This proposal was to have a 65-hp Austro-Daimler engine. Allocated to Eastchurch it was not delivered. No Short c/n is known for this machine.

Short Dual Control Seaplane. RNAS Serial No. 182

Ordered under Contract CP46634/14 this single machine was on order for Grain in the second half of 1914, but never delivered.

Short Folder Torpedo Carrying Tractor Seaplane. RNAS Serial No. 186

This twin-engine machine was ordered in July 1914, under

The Spithead Royal Naval Review 18 to 22 July 1914

This occasion saw the Naval aircraft put on a display to show that aircraft were part of the Fleet. It was covered in detail in the aviation press of the day.

During the week previous, several flights had assembled at Calshot. One from the Isle of Grain, one from Dundee, another from Felixstowe, one from Yarmouth in addition to that of the Calshot station. The machines had to fly out on Saturday morning to their special moorings off Fort Monckton, in the entrance to Portsmouth Harbour, in order to be in position for the review.

The first to start were four 160-hp Gnôme Short Folders from the Isle of Grain, under the command of Sqn Cdr Seddon (Lt, RN). These machines had their vertical exhaust stacks painted with white rings: one to four rings, to identify the four Folders. There was trouble in *getting all the engines to start. One gathers,* commented *The Aeroplane, that these "double-row" engines – i.e., two rows of seven cylinders each – are by no means as satisfactory as others.* The vertical 'chimney' exhausts had been fitted to overcome the problems with these engines overheating. Samson had his old No. 10 fitted with a similar engine in order to experiment with it.

The next flight to get away was from Dundee under the command of Sqn Cdr Gordon (Maj, RMLI). This flight consisted of Short tractor biplanes with the nine-cylinder 100-hp Gnôme. One machine cracked an airscrew in getting off, and had to have a new one fitted.

Felixstowe sent Maurice Farman seaplanes, there with 100-hp Renault engines and one with a 130-hp Salmson engine, under Sq Cdr Risk (Capt, RMLI).

The Yarmouth flight under Sqn Cdr Courtney (Lt, RN), were on 120-hp Gnôme powered Henri Farmans.

Sq Ldr Longmore (Lt-Cdr, RN) was to lead the Calshot flight, however *the Sopwith gun-carrier (200-h.p. Salmson) which the commanding officer was to have flown, refused to get off the water, chiefly owing to a defective propeller, but partly because an experimental alteration had been made to the tail and partly because the engine took a fit of sulks, and there was no time to get things right.* Longmore had organized the accommodation at Calshot for the other flights and had superintended their departure that morning only to be deprived of starting at all. *Consequently, this flight was under Flight Commander Bigsworth, who started first on the Short gun-carrier (160-h.p. Gnome), a huge machine which seems able to lift anything, but otherwise does not seem to fly as well as the Short tractors, and appears to me to need more surface aft to counteract the big projecting nacelle and gun. ... The gun on the Short is a 1½-pounder, more or less like to the pom-poms used by the Army some time ago. It is the biggest weapon yet used in an aircraft. It was first used on the Sopwith, and later was used to test the Shorts ability to stand the recoil.*[1] The next machine was the Sopwith 'Circuit of Britain' seaplane, while the lasts machine was a flying boat, the Sopwith bat-boat.

After the last warship had passed the Royal yacht, on receipt if a wireless message from a Maurice Farman scout, seventeen aircraft in single-file took part in a fly past, dipping as they flew over the Royal yacht. Two came down with engine trouble and were towed back by torpedo boats. After this three Navy land machines from Eastchurch flew over.

Flight noted that *the sight of seaplanes flying over the King's yacht, dipping in salute, as they flew and manoeuvred in the air, seems to have impressed people with a feeling that in these craft we have the potential of a development in war as great, if not even greater, than those bound up in the submarine.*[2]

Notes:
1. 'The Royal Naval Air Service in Being.' Signed C.G.G. (C.G. Grey). *The Aeroplane*, June 22, 1914. P. 79.
2. 'Aircraft at the Naval Review.' In Editorial Comment. *Flight*, 24 July 1914. P. 773.

Contract CP 02112/14, for Calshot. Power was variously referred to as two 200-hp Salmson, 160-hp Le Rhône or 180-hp Gnome engines. It is thought that the ideas of Samson that a two-engine machine was required to carry a torpedo would have been behind the design. On order to at least October 1914, it was never delivered to Calshot as contracted for.

Two further Short Admiralty 186 Type folders were ordered and allocated RNAS serial Nos. 1448 and 1449. Power was to be two 200-hp Canton Unné engines.

These two are listed in the 'Progress of Seaplanes building shewing probable delays and deliveries' chart for 27 March 1916, with the note 'Postponed.' They are last listed on the Report for 26 June.[104]

Short S.96 to S.105 The Russian Seaplanes

Research by Gordon Bruce has found that the missing Short construction numbers S.96 to S.105 were listed in the Shorts' Order Book as ten 135-hp Seaplanes for the Russian Navy.

The *Times* of 20 March 1914, contained an advertisement by the Imperial Russian Ministry of Marine Institute for a 'Competition in Designs of Seaplanes.' The following prizes were offered for *designs adjudged of sufficient merit*:

1st Prize: 3,000 roubles (about £318)
2nd prize 2,000 roubles (about £212)
3rd prize 1,000 roubles (about £106)

Details of the specification could be obtained from the

Russian Naval Attaché, South Kensington, SW.

The *Times* followed with an article on 23 may 1914 titled 'Seaplanes for Russia.' This set out the specifications as follows: - Crew of five (770 lb), a load of 320 lb, exclusive of fuel, oil and water. Range with this load to be 630 miles at not less than 63 mph. The number of motors not less than two. Horizontal flight to be available with two-thirds available horse-power. Motors to be accessible in flight. Provision for two pilots and they were to be able to relieve each other in flight. Provision for a clear field of vision for observers and for mounting a machine-gun, wireless, searchlight and bomb-dropping apparatus.

Who applied for one of the prizes, if indeed anyone did, is unknown, but the capabilities of contemporary aircraft were not up to the stated requirements? Clearly the proposed Short 135-hp machines were not to this specification.

S.90 to S.95 were allocated to a January 1914 order and were Nos. 161 to 166, 200-hp Canton-Unné seaplanes.

S.106 to S.107 were allocated to the prototypes of the 225-hp Sunbeam Short 184 Type Nos. 184 and 185, ordered in July 1914.[105]

This shows that S.96 to S.105 were issued pre-war between January and July 1914. The order was divided into: -

S.96 to S.100. Tractor seaplanes.

S.101 to S.105. Gun carrying seaplanes.

The 135-hp engine would probably have been the single-row Salmson Canton-Unné. Gordon Bruce suggests that S.96 to S.100 would have been equivalent to the Short No. 135 patrol seaplane. S.101 to S.105 were to be fitted with a Maxim machine gun rather than a quick-firing gun. What, if any, of the existing Short gun carrying machines, No. 66 and No. 126, this machine would have resembled is lost to history.

Why the Russian order was not fulfilled is unknown? The Admiralty could have persuaded Shorts to give up the contract on the outbreak of war, or more likely used the powers of the Defence of the Realm (Consolidation) Act of November 1914, to reserve the output of Shorts to itself.

Endnotes

1. This undated paper was written by the then Air Commodore Samson after his retirement. Copy in IWM via G Bruce.
2. Driver, H. *The Birth of Military Aviation – Britain, 1903-1914,* The Royal Historical Society, UK, 1997.P.53.
3. Jarrett, P. *Frank McClean, Godfather to British Naval Aviation*, Seaforth Publishing, UK, 2011. PP. 59 & 65.
4. See Sturtivant, R & Page, G. *Royal Navy Aircraft Serials and Units*, Air Britain, UK, 1992. P.19.
5. Notes on Wing Cdr Stanley T Freeman. RAFM, J.M. Bruce Collection Box 24.
6. For details of the training of the Territorials see Jarrett, *Op Cit*. P. 75. The War Office stopped their training in February 1912. *Nevertheless, four members had made enough progress to obtain their certificates. At a subsequent date others qualified.* Turner, C.C. *The Old Flying Days,* Samson & Low, Marston & Co Ltd, UK. 1927. P.20.
7. Although the catalogue states that an 80-hp engine could be fitted, the specification table only records the 70-hp Gnôme.
8. Samson's Log Book in IWM.
9. 'Naval and Military Aeronautics.' *The Aeroplane*, 25 December 1913, P.680.
10. Bruce, J.M. *The Aeroplanes of the Royal Flying Corps – Military Wing,* Putnam, UK, 1992. P.484.
11. The story related by Harold Penrose in his *British Aviation – The Pioneer Years*, (Putnam, UK, 1967. P.345), that the T5 was the machine built from Grace's plans is incorrect. This is a fault in relying on oral history without checking original documentation.
12. Captain Grahame Douglas – Experiences. IWM Sound Archive. Accession No. 000018/11. P.18-19.
13. Short, O. 'Twenty-one Years' Progress of Seaplane Design,' *Flight,* 3 January 1930, P.67.
14. *Ibid.* P.68.
15. 'The First Flight of the First Seaplane.' *Evening News,* 7 September 1919. In this interview, Oswald states that due to the requirement for volume production he *was not allowed to experiment on new ideas. We therefore concentrated on improving existing models, and were not able to put into effect our ideas on all-metal seaplanes.* As the machine could not take-off from the water, the article's title is a misnomer.
16. 'History of Naval Aeroplanes and Seaplanes' by Colonel C.R. Samson, *Op Cit.* In another paper Samson states that the deck used wooden supports for cheapness and the size of everything was increased *in order to make a solid proper job of it,* which handicapped him as it was not as portable as it should have been. 'Early British Naval Aviation.' Undated paper in IWM via G Bruce.
17. *Ibid.* In this document Samson states that he flew off this platform in December 1911.
18. Turner, C.C. *The Old Flying Days,* Samson & Low, Marston & Co Ltd, UK. 1927. P.23. Turner has been called the first aeronautical correspondent and was on hand reporting on most of the early activities taking place in the 'Old Flying Days.'
19. 'Early British Naval Aviation' Undated paper by C.R. Samson in IWM via G Bruce.
20. The official *Naval and Military Aeronautics – 1913* report gives the speed as 10½ knots, the same as does Samson. TNA AIR1/2126.
21. 'Early British Naval Aviation.' *Op Cit.*
22. Quoted in Sturtivant & Page. Sturtivant, R & G Page, G. *Royal Navy Aircraft Serials and Units*, Air Britain, UK, 1992. P.19.
23. Samson's Log Book in IWM.
24. 'Developments at Eastchurch.' *The Aeroplane,* 17 October 1912. P. 395.
25. Lt Wilfred Parke's obituary appears in *The Aeroplane,* 19 December 1912. P.616. Parke was killed while flying

a 70-hp Handley-Page monoplane from Hendon to Oxford with Arkell Hardwick, Handley-Page's chief business assistant as passenger. Hardwick was still breathing when pulled from the wreck at the Wembley Golf Club, but died soon after.

26. Sorley, R. 'Sorties with Sorley' *Cross & Cockade International*, Vol.37 No.2, P.120.
27. Jarett, P. *Op Cit.* P.108. G Bruce has found no evidence to show that the rebuilt S.32 ever flew as a seaplane. E-mail to author 13.07.2020.
28. 'A Visit to Eastchurch.' *The Aeroplane,* 14 August 1913. P. 188. Two photographs of the rebuilt No. 3 grace the article.
29. *It arose from the fact that money for new aircraft had to appear in the Naval Estimates as a separate item, whereas a lump sum covered unspecified repairs. When an old aircraft was returned for repair, the contractor was enabled to design an entirely new aircraft, though he was supposed to incorporate something of the old one.* R Bell Davies, *Sailor in the Air,* Peter Davies, UK, 1967. P.75.
30. C.R. Samson Log Book – 'Flights made by Lieutenant CR Samson RN.' Original in IWM.
31. 'Naval and Military Aeronautics.' *The Aeroplane*, 30 October 1913. P.478.
32. 'Naval and Military Aeronautics.' *The Aeroplane*, 25 December 1913. P.680.
33. 'A New Short Biplane.' *The Aeroplane*. 28 November 1912. P.550, & 'A Correction to be Noted'. 5 December 1912. P.573.
34. 'The Week's Work.' *The Aeroplane,* 14 November 1913. P. 300. These entries in 'The Week's Work' give the serial numbers and crews of the military aircraft and their flights. They are full of the names of pioneers in British Army and Naval aviation.
35. Weekly Report Sheets P.17, signed by Carpenter J.V. Brownrige(?). Copy in RAF Museum J.M. Bruce Collection Box 5.
36. 'War-Time Reminiscences' by Flt Cdr A.H Sandwell. *Canadian Aviation,* Official Journal of the Aviation League of Canada. These articles were reproduced in *Cross & Cockade Great Britain,* Vol.13 No. 1, 1982.
37. [37] 'An Armed Naval Biplane,' in 'British Notes of the Week'. 'Royal Aero Club Eastchurch Flying Grounds,' in 'From the British Flying Grounds.' *Flight,* 1 Nov 1913. P.1203. From the articles it appears that Gordon Bell's flight was made on Wednesday 29 October 1913.
38. Vernon Hillier, A.H. *Brief History of the early Short "S.27" Type Aircraft & subsequent variants. Short S.26 to S.62 inclusive.* FAA Museum, June 1980.
39. Samsons Log Book in IWM.
40. Sorley, R. 'Sorties with Sorley' *Cross & Cockade International*, Vol. 37 No. 2, UK, 2006. P.121.
41. 'Naval and Military Aeronautics', *The Aeroplane*, 15 July 1914. P.62.
42. Jarrett. P. *Op Cit.* P. 116.
43. Bruce, J.M. *The Aeroplanes of the Royal Flying Corps – Military Wing,* Putnam, UK, 1992.
44. Weekly progress report of Admiralty Air Department 'I' Section for W/E 18 September 1915. TNA AIR1/2234/15/226/56.
45. *Ibid.*
46. McMahon, L & Partridge, M. *A History of the Eastbourne Aviation Company 1911–1924.* Eastbourne Local History Society, UK, 2000. P.102.
47. Cole, C & Chessman, E.F. *The Air Defence of Britain 1914–1918,* Putnam, UK. 1984. P.127.
48. Hand written notes on interview with Wing Cdr Stanley T Freeman. The pilot was possibly Hewlett who taught Freeman to fly. RAFM, J.M. Bruce Collection Box 24.
49. 'Early British Naval Aviation.' This undated paper was written by the then Air Commodore Samson after his sretirement. Copy in IWM via G Bruce.
50. Report by Samson. Notes from TNA AIR1/626/17/88 by J.M. Bruce. RAF Museum J.M. Bruce Collection Box 3.
51. Sturtivent, R & Page, G (*Royal Navy Aircraft Serials and Units 1911-1919,* Air Britain Publication, UK, 1992) note that it has been suggested that Nos. 423 & 424 were transferred to the RNAS as Nos. 1268 & 1269, but on present evidence this seems unlikely.
52. Bruce, J.M. *The Aeroplanes of the Royal Flying Corps – Military Wing,* Putnam, UK, 1992.
53. L'Estrange Malone became the first Communist member of the UK Parliament
54. Quoted in Bruce, J.M. 'The Short Seaplanes.' *Flight* 14 December 1956, P.923. S.47 is confirmed by Job No. 570: *fixing wireless installation in Biplane No 4,* Shorts' Order Book entry of 23 October 1912.
55. 'Air Eddies.' *Flight,* 2 March 1913. P.192.
56. 'The Short Monoplane.' *Flight*, 9 March 1912. P.212.
57. Samson's Log. IWM. Copy via Gordon Bruce.
58. Minute, 27 February 1912. Notes from TNA AIR1/626/17/88 by J.M. Bruce. RAF Museum J.M. Bruce Collection Box 3.
59. R Bell Davies, *Sailor in the Air,* Peter Davies, UK, 1967. P.72.
60. Samson's Log Book entries. Books in IWM.
61. Lankester Parker, J. 'The First Short Brothers Commemorative Lecture'. Reprint from *the Journal of the RAeS*, Vol. 66, No. 63, January 1962. Bell Davies recorded that it was known as the 'Double Dirty'. *She was a monoplane of immense span with one tractor engine and one pusher. The pilot sat between them in a constant bath of black lubricating oil.* R Bell Davies, *Op Cit.* P.72.
62. Notes from interview with W/Cdr S.T. Freeman. 22 August 1964. RAFM J.M. Bruce Collection Box 24.
63. Although shown as a landplane in most printed histories, contemporary references indicate it was an amphibian. Letter to author from G Bruce. 12.10.20.
64. Samson's Log Books in IWM.
65. 'Developments at Eastchurch.' *The Aeroplane*, 17 October 1912. P. 396.
66. Notes Briggs' flying log books via G Bruce. Samson's Log Books. Both in IWM.
67. 18 January 1913 entry. IWM. Notes from Briggs Log Book via G Bruce.
68. Quoted in Empson, R.H. 'Eastchurch: The Early Years.'

Reprint of Royal Aero Club history in *Cross & Cockade International*, Vol.40 No. 2, 2009. P.77.

69. Sturtivant & Page. P.19. According to Barnes, Samson made the first flight on 21 October 1912; however, the statement in *The Aeroplane* indicates that it was flown before this date. For a more detailed account of the S.46 see Jarrett, P. *Op Cit.* P. 104.
70. Travers, E. *Cross Country,* Hothersall & Travers, UK. 1989. P. 66.
71. 'The Week's Work.' *The Aeroplane,* 1 August 1912. P.117. McClean realized that a single-engine machine was not sufficiently safe to operate over water and for reliable over-water flying multiple engines were needed, hence his order for a 'Multiple Plane Machine.' Jarrett, P. *Op Cit.* P. 59.
72. See Phil Jarret's notes in *Cross & Cockade International,* Vol. 40, Autumn 2009. P.204.
73. Jarrett, P. *Frank McClean, Godfather to British Naval Aviation,* Seaforth Publishing, UK, 2011. Gordon Bruce points out that the trials of the S.40 with pontoon floats took place at Harvey Ferry that was isolated while his flight down the Thames was very public. Letter to author 19.09.2020.
74. *Flight,* 17 August 1912, P.759.
75. Jarrett, P. *Op Cit.* P. 177.
76. Gerdessen, F & Geldhof, N. *De Internneringen van vliegtuigen tijdens De Groote Oorlog,* Uitgeveri Geromy bv, Netherlands, 2018.
77. *The Aeroplane.* 13 Feb 1913. P.179. Percy Grace was the brother of the late Cecil Grace and was an agent for Shorts.
78. 'The Olympia Exhibition 1913.' 'Short Machine (Mr. Percy Grace).' *Flight* 22 February 1913. P.222.
79. "Naval and Military Aeronautics', *The Aeroplane* 7 August 1913. P.152.
80. *The Aeroplane,* 29 January 1914. 'Naval and Military Aeronautics.' P.107.
81. These notes are taken from the 'Naval and Military Aeronautics' column from *The Aeroplane* in 1914. This column gives a day-by-day report on the activities of the military aircraft in the UK with reference to serial numbers and crews. The column for 9 April criticized the decision to establish a permanent base at Dundee. *Apparently Leven is the best place for manoeuvres, but no good as a base.* The correspondent stating that the decision to set up at Dundee was due to a decision of Mr Churchill, and *his constituents would be gratified by his decision.*
82. The SS *Empress* had been requisitioned and commissioned on 11 August 1914.
83. Samson, C.R. *Fights and Flights,* Ernest Benn Ltd, UK, 1930. P.3.
84. *Ibid.* P.48.
85. *Ibid.* P.76.
86. 'The "Daily Mail" £5,000 Prize.' *The Aeroplane,* 3 July 1913. P. 19.
87. 'The Short "Circuit" Waterplane.' *The Aeroplane,* 31 July 1913. P. 128.
88. Jarrett, P. *Op Cit.* P.123.
89. Sandwell, A.H. *Op Cit.*
90. Henry Meyrick Cave-Brown-Cave had a remarkable career in the RNAS and RAF. He was the OC of the Far Eastern Flight that flew Supermarine Southampton flying boats to Australia in 1928. His active career came to an end when he was seriously injured following an aerial accident on 17 January 1939.
91. Moore, W.G. *Early Bird,* Putnam, UK, 1963, P.11. The only known photographs of No. 80 are reproduced in this book.
92. 'From the British Flying Grounds.' *Flight,* 24 July 1914. P.785.
93. McClean tried to give the machine to the Navy at the beginning of the war but the Admiralty insisted on paying him £1,200. Mason, T. 'Royal Naval Air Service Aeroplanes of the Royal Naval Flying Corps (Naval Wing) and RNAS 1911 to 31 August 1915.' *Cross & Cockade Great Britain,* Vol. 15, No. 4, 1984. P.155.
94. TNA CAB 4/5/2/179B.
95. The Vickers was the first of the pom-pom guns to be ordered for the Royal Navy. It was a scaled-up Maxim machine gun that fired belt fed explosive rounds and was most probably 37 mm calibre. The name pom-pom came from the relative low rate of fire of the explosive rounds. Gordon Bruce research, letter of July 2020. The late J.M. Bruce described the weapon as a *Fearsome bit of ordnance.* The family of this type of weapon can be seen on Forgotten Weapons.com.
96. Extract from the Archives of Sir Arthur Longmore. *Manuscript Log of RNAS Calshot 6 May – 24 September 1914.* Copy in FAA Museum.
97. Post-war recollections of early bomb and armament experiments. Dated 25.11.1920. TNA AIR1/725/118/1.
98. Details of the tests are contained in the following files: TNA AIR1/189/15/226/2 & TNA AIR1/189/15/226/6.
99. TNA AIR1/725/118/1. The 27 mm guns saw little service during the war but were sometimes fitted to the large flying boats of the RAF such as the Short Cromarty, and the WWII Sunderland was originally specified with a 37 mm gun. A Short-built Felixstowe F.5 for Japan was fitted with a 1 pounder gun.
100. Quoted in Penrose, H. *British Aviation – The Pioneer Years,* Putnam, UK, 1967. P.490. Camden Pratt wrote of Short Number 0/1378. The S.81 was Job No. 1378. He must have had a prodigious memory!
101. Grey, C.G. *Sea Flyers,* Faber & Faber, UK, 1942. P.78.
102. The Davis gun fired a load of buckshot out the end of the tube that was the gun's barrel. This counterbalanced the weight of the shell fired forward, eliminating recoil. It was tried by the British and the USN, the latter even building a special Davis gun carrier biplane, but it was never very successful.
103. Snowden Gamble, C.F. *The Story of a North Sea Air Station,* University of Oxford and London, UK. 1928. P.124.
104. TNA AIR1/152/15/119.
105. Sturtivant & Page. *Op Cit.*

Chapter 3: Short Tractor Hydro-Aeroplane Type S.41

Samson wrote that at the end of 1911, *Mr. Horace Short and myself (I having obtained permission from the Admiralty) were working out* (the) *design for a seaplane. We considered (1) Flying Boats (2) Floats, and we decided on building a float type.*[1] *We were faced with, what for in those days were real difficulties. No one knew anything about float shapes, planning surfaces, steps, and such like important matters.* They decided on the two main float hydroplane arrangement. Horace placed the floats well forward *to avoid any chance of diving* and allowed a buoyancy of twice the fully loaded weight. In order to obtain strength with lightness, the floats were of wooden framework with internal bracing wires, ply sides and bottom, with fabric top surface.[2]

Oswald Short wrote that the first seaplanes were produced in the latter half of 1911 and during 1912 by the process of replacing the wheels and skids of aeroplanes by flotation gear. This development was approached with great caution by the experimenters of the day. The effect of alighting on the water at high speed was an unknown quantity. It was felt that a couple could be set up that would cause the machine to turn over in spite of the use of the elevators. Short Brothers *produced a twin-float tractor seaplane with wing-tip floats, and in addition a tail float which gave a three-point suspension to the machine on the water. The main floats were flat-bottomed without a step, but they were places in such a position that the rear edge of the floats acted as a step in breaking up the continuity of the water flow at a suitable distance behind the centre of gravity.*

The shape of these floats was crude in the extreme from the point of view of air resistance, but the flat under-surface of the floats was highly efficient in its planning properties, and made for success in permitting the machine to take off with the low horse-power then available in comparison with the all-up weight of the machine. This type of machine became almost universally used and lasted without radical change in appearance until as late as 1920.[3] The machine that Oswald was thinking of was the S.41, photographs of which illustrated the article.

Short S 41 – Tractor Biplane Hydro Aeroplane RNAS No. H1/10

The S.41 and S.45 were generally the same as the S.36 and both were purchased by the Admiralty. They both had land and float undercarriages. The Shorts' Order Book shows that the instructions to build the S.41 were issued on 19 January 1912.

The rectangular fuselage of the S.41 was suspended above the lower wing. The aircraft had tandem seating with the pilot in front. The observer's cockpit could accommodate two if necessary. Both upper and lower wing centre sections were uncovered on the prototype machine. The wings had extensions to the upper panels that were braced by long sloping struts from the bottom of the outer interplane struts. Ailerons were only fitted to the upper wing. When the machine was not in flight the ailerons hung down as there were no balancing cables fitted. The rectangular rudder had no fin but was braced by a strut running from the top of the rudder to the fuselage.

The Short Type S 41 is described in a Short Brothers catalogue as a Tractor Hydro-Aeroplane for naval reconnaissance work. The machine could be *fitted with either 80, 100, or 160 H.P. "Gnome" engines, or can be adapted to take stationary engines, if required.* With a 100-hp Gnôme it could carry a pilot and up to two passengers, with four hours fuel, wireless telegraphy apparatus, etc., *and a self-starting arrangement by which the engine can be started up by the pilot without having to leave his place.*

Wheel control if fitted for warping and elevating, the rudder steering being operated by foot-bar.

Both pilot and passenger are provided with comfortable roomy seats, fitted with light air cushions. The seats are tandem fashion, and so placed as to give an excellent view for reconnaissance work. The pilot faces a roomy dashboard containing all necessary instruments and switches, petrol control, etc., in the most convenient position for operating.

The machine throughout is constructed in an exceptionally strong manner, being designed to withstand the battering effect of waves when at sea. It is very stable both on water and in the air, is quick in getting off the water and is excellent in alighting on a rough sea.

The planes and fuselage are covered with a special fabric which is perfectly water-tight, keeps a permanently taut and highly-polished surface, and is not affected by changes in temperature.

The float chassis is of exceptionally strong and reliable manufacture, designed to withstand heavy seas, when the machine is floating. The two floats are of special construction, containing a number of water-tight compartments, and have a large reserve of floating power.

The engine can be started from the pilot seat, and the engine bonnet is so arranged as to give perfect accessibility to the engine and attendant parts.

Steel fittings are used throughout in its construction, all fittings possessing a large margin of safety over actual working strains.

These machines can be fitted with wireless telegraphy if desired.[4]

Right: RNAS No. 10 taxying. Note the open centre section.

Below Right: Close-up of the engine of the S.41. The engine mounting with front cross-bearer and the smaller gap between the fuselage and the lower wing were the only visible difference to the S.36 in front elevation.

Below: Side elevation of the early S.41/RNAS No. 10.

The S 41 (H1/No. 10) had first appeared early in 1912, Commander Samson flying it on 27 March with a landplane undercarriage. *I found she flew quite well.* Samson recalled that he *was fortunate enough to get the Admiralty to buy one of the first 100 h.p. Gnome engines.* Further test flights were made on 2, 10, 12 and 13 April, mainly to find the correct type of airscrew, *amongst other things.* He flew it to Grain on 25 April where the floats were fitted. *On April 26th I tested her on the water and found that the tail float was fixed too high out of the water. April 29th, everything was ready for further trial, but neither time nor space permitted of anything except a fast taxi.*[5] Air bags on the wing tips provided stabilization on the water.

With no time left for more testing, S.41 was embarked on board HMS *Hibernia* for the King's review of the Fleet.

Left: Samson in his H.1/ No. 10 were the centre of attention wherever he stopped.

Below: Drawing of the S.41 from the Shorts catalogue. This depicts the machine with the open centre section.

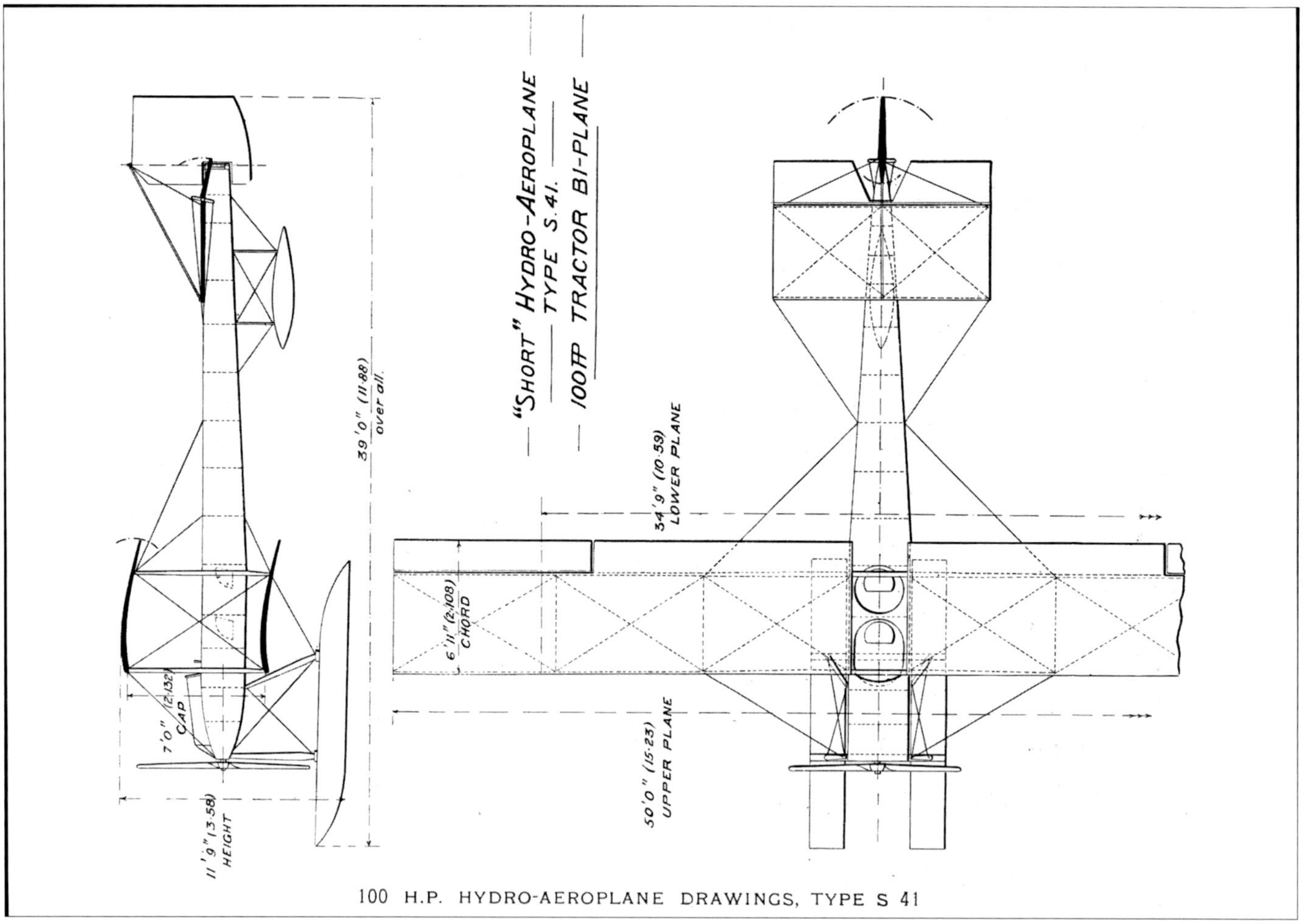

On 3 May, Samson *took No.10 off the water and safely alighted after a 10 minute flight. This flight demonstrated the first practical seaplane.* These flights showed that Naval aviation was practical and from that time Samson got *practical support.*[6]

As noted above S.41/H1 was fitted with floats for sea trials on the 26th, however, the trials failed as the floats were rigged incorrectly. The next trial on the 29th again failed due to the weather, the H1 being driven ashore. Taken aboard *Hibernia* and delivered to Weymouth for the Fleet Review, the H1 was taken ashore by lighter as

Right: No. 10 with land undercarriage, rounded lower wing tips and revised outer wing bracing.

Left: Samson in the S.41 that became his beloved No. 10. The machine was the first British naval aircraft to mount a machine gun. Frank Susans is in the rear with the Maxim machine gun.

Right: The S.41 with experimental folding wings. Eastchurch, November 1913. 140-hp Gnome engine.

Left: No. 10, with straight wingtips to all wings, surrounded by Royal Marine personnel at Marske or Redcar.

Left & Above: Wing Cdr Gerrard, the CO of No. 2 Wing, inspecting No. 10 after its conversion. Imbros, 1915. The S.41 was reconstructed by the makers in August 1915 with a fuselage similar to the Improved Type 74, and with a 135 hp Canton-Unné engine. The fuselage insignia is the Union flag and it carries the red/white RNAS rings on the lower wings.

Right: The S.41 in a still later scheme with rudder stripes and the serial on a white rectangle on the fuselage side. The wing rings have a white outline.

Right: Sea water pours from the broken wing of the landplane conversion of No. 10 as she is hauled aboard HMS *Grafton*. The red ring marking has a white outline.

Below: The sad end of No. 10, burnt on Imbros after salvaging.

a landplane. From photographs showing the H1 being landed at Weymouth as a landplane, if can be seen that the slipway at Weymouth was too steep for the H1 to be safely landed with the floats projecting beyond the forward edge of the lighter. The decision was probably made to land it in landplane configuration and then fit the floats once ashore.[7] The wooden pontoons of these floatplanes were fragile and beaching had to be handled with extreme care. The first flight as a seaplane is described above.

Samson recalled that it *was in 1912 that we made our first seaplane, known first as Hydroaeroplane H 1 and later as Short No.10... She was an historic machine. I used her from March 1912 till the war came and never broke a piece of wood in her. I flew her in the Army manoeuvres of 1912, took her to Scotland for submarine experiments at Scapa Flow, used her for similar work at Harwich, and made the seaplane duration flight with her from Sheerness to Portsmouth. She was the first machine to have folding wings which were fitted and the first from which a machine gun was fired.*[8]

Converted back to a landplane, it took part in the September 1912 Army Manoeuvres.[9] In January 1913 it mounted a Maxim gun for experiments.[10] This converting back and forth appears to have been a feature of S.41's service life.

S.41 was the first British naval aircraft to be armed with a machine gun. The gun was a 0.45 in Maxim.[11] Samson's flying log (in the IWM) shows that on 10 February 1913, he piloted No. 10, as H1/S.41 had been restyled after the Weymouth incident, with Engine Room Artificer (ERA) Susans as passenger. When *firing a Maxim gun with ball amm,* (Susans) *felt no shock all.* (*Ball* was Service speak for a solid small arms slug even though it is cylindrical with a pointed end.)[12]

Lt Robert H Clark-Hall also flew with Samson on gun firing trials. He reported on the trials of guns from airships and aircraft, and noted that he ordinary 303 maxim weighed 60 lbs were used but that lighter versions were being trialled at the Navy's gunnery establishment. *A special Maxim carrying aeroplane is now at Eastchurch and a series of firing experiments are in progress. In this machine both the ordinary Maxim and a lighter pattern by Vickers are being tried.*[13]

No. 10 was probably the machine that Group Capt R.H. Clark-Hall, RAF, was thinking about when he wrote in 1920 that *An ordinary service Maxim gun...was mounted on a pillar in the observer's seat of a tractor machine, and firing carried out which revealed the difficulties to be overcome, not only of aiming appliances and mountings, but in manipulation of belts in the high air speeds, and the provision of safe methods of ejecting the empty cylinders.*[14]

As No.10 the machine underwent many modifications including the fitting of larger folding wings.[15] Early in 1914 a 140-hp Gnôme was fitted to *Commdr. Samson's old favourite*[16] replacing the 100-hp Gnôme. It remained at Eastchurch when 2 Wing went to France after the declaration of war. Completely rebuilt by its makers in August 1915, it emerged resembling an Improved Type 74 with a 135-hp Canton-Unné engine. This was probably the reported type 140-hp engine it had on 27 June 1915, when it was included in an armament report at Eastchurch. The report on No. 10 gave the following details: - The pilot arrived with automatic pistol. A Dropper was fitted for 14 J Grenades. It was in the Aegean by October 1915. Fate unknown but probably lost in the fire of 4 November 1915.

Short S.56 & S.57 – Tractor Biplane Hydro Aeroplane RNAS Nos. 20 & 21

Nos. 20 and 21 were ordered under Contract C.P.43037/13 as Improved Type S.41 tractor seaplanes. They were initially to have the serials H8 and H9 but were given serials Nos. 20 and 21 instead. Apart from the 100-hp Gnôme engine, they in no way resembled the S.41. They were similar in appearance to S.42. While machines could vary markedly from the prototypes in those days before standardisation became the norm, why these two were called the Improved Type S.41 is unknown.

A balanced rudder with triangular fin was fitted with the tailplane mounted on the top longerons. Twin wooded pontoons were mounted with a small stabilizing float at each lower wingtip. A tail float was also fitted. The upper wing extensions were cable braced from kingposts. The centre sections were covered and longer ailerons were fitted.

Nos. 20 and 21 were both delivered to Grain by Gordon Bell[17] on 22 April 1913.

Wireless telegraphy was experimented with at the Isle of Grain by Lt Raymond Fitzmaurice and in May 1913 he tested an installation in Short No. 20. *She was finished in time to carry out W/T trials and take part in escorting the Royal Yacht the last 50 miles of her journey back from Flushing to Port Victoria after the German wedding. We proceeded out to sea in the morning and flew out in the direction of Flushing and picked up the Royal Yacht about 20 miles to seaward of the Foreland. We gave them a W/T salute, the first W/T message the King had ever received from aircraft, I think. The apparatus worked excellently, and our base at Grain had us in good touch all the time we were away, the distance being about 50 miles.*[18]

No. 20 saw service at Great Yarmouth and was recorded by C.F. Snowden Gamble in his classic *The Story of a North Sea Air Station.* Referring to the progress made at Great Yarmouth with permanent buildings in July 1913, with pilots and new machines arriving he recalled: -

Of the seaplanes, one was a 200 b.h.p. Gnome-engined Short, a float machine with wing-tip floats under the lower plane and also a small tail float with a water rudder. All the floats had watertight compartments. This machine had two seats, accommodation for two persons.

Small stabilizing floats were fitted at the lower wing tips. While the fuselage was almost identical to the original S 41, the wings were different. No. 20 had cable bracing from king-posts supporting the upper wing extensions. The centre sections were covered and longer ailerons were fitted.

The Royal Navy Manœuvres of 1913

The 1913 Naval Manoeuvres were intended to simulate the defence of the East Coast (Blue fleet) against an invading naval force (Red fleet). There was also a trade protection exercise carried out off the west coast of Ireland. Little imagination was required to realise that the Red fleet was meant to represent Germany. A large part of the Home and Atlantic Fleets were involved; 16 dreadnoughts and 24 pre-dreadnoughts, 5 battlecruisers, 58 cruisers and 139 destroyers, plus many smaller craft and submarines. Included in the total were eight seaplanes and HMS *Hermes*. The Report by the Umpire-in-Chief, Admiral of the Fleet Sir William Henry May, makes for dry reading but includes a separate Appendix covering the work of the seaplanes *as it was not possible to embody it without confusing the narrative, and also because this is the first time aircraft have taken part in the Manoeuvres on the high seas.*[19] The fleet manoeuvres commenced on 23 July 1913 and were terminated by Admiralty order on the evening of 1 August.

The Blue fleet had air stations at Cromarty and Leven with three floatplanes at each base but owing to the docker's strike at Hull, some of these machines arrived late. The Red fleet had *Hermes* with two floatplanes and an air station at Great Yarmouth with three seaplanes and one aeroplane in service:

No. 20. Short seaplane equipped with W/T.
No, 29. Maurice Farman seaplane.
No. 84. Borel seaplane.
No. 67. Maurice Farman aeroplane.

Despite difficulties operating from extemporised bases, off unprotected beaches, all the land based machines were able to complete a number of patrol and reconnaissance flights. Submarines and other vessels were observed and reported but could not be attacked as the rules defined their role to be entirely that of reconnaissance. These were a foretaste of the unending and frequently tedious patrols flown from shore bases throughout the First World War.

Hermes, based at Great Yarmouth during the manoeuvres, carried two floatplanes; Borel monoplane No. 48, and Short Admiralty Type 74 Folder Tractor Seaplane, No. 81, an improved version of Samson's 'Amphibian'. It had been built with folding wings that reduced its 56 foot wing span to only 12 feet, greatly reducing the required on-board space for storage. It was also fitted with an early Rouzet radio transmitter, but no receiver. Commander Samson was aboard *Hermes* with Lt Raymond Fitzmaurice, RN, to act as observer and radio operator. As with the land based machines, *Hermes* and her floatplanes were considered to be immune from attack, and were not permitted to make attacks themselves. *Hermes* and her machines were intended to represent an airship attached to the Red fleet, as such she was permitted to proceed up to 300 miles from Great Yarmouth, but must return within 48 hours.

On 21 July, *Hermes* launched Samson in Short Folder No. 81 in a moderate sea. He made a successful flight and repeated this in the evening. *On each occasion she got off the water well.* No. 48 had difficulty in getting off the water and *the engine was shifted and* (the) *machine again put in flying trim.* The following day no flying was attempted due to the weather. War was to commence on the morning of the 23rd, however *nearly a full gale was experienced.* This tested the housing for the aircraft. Borel No. 48, that was stowed on the forward platform, had her right wing completely broken off. No. 81, carried aft, was slightly damaged when the wind caused the canvas hangar to collapse on the aircraft but was reported as having *stood the wind well and was little the worse for the heavy rain.*

Caudron G.II, No.55, an amphibian with wheels partially buried in its floats, replaced the Borel on the forward platform. Its sole contribution to the manoeuvres came on 28 July when, with *Hermes* steaming at 10 knots into a light wind, the Caudron successfully launched from the platform and *The* Seaplane *rose easily and made a good flight into Yarmouth.*[20] The pilot is unknown, but as Samson is the only pilot known to be attached to *Hermes,* it would seem that that he made the flight, although it is not recorded in his Log Book.[21]

The bulk of flying from *Hermes* fell to Short No. 81, with Samson and Fitzmaurice as crew. Over the ten days of the manoeuvres the Short made seven flights. Radio transmissions were received by Hermes during many of the flights. On the 24th No. 81 made a brief reconnaissance during which she was in contact with the carrier by W/T. No. 20 was also active on this day, carrying out a flight early in the morning from the air station to the Leman Bank and returned via the Haisborough light vessel. *Owing to the heavy surf, great difficulty was experienced in beaching the machine after she alighted, but the task was accomplished successfully later.* Later the land machine No. 67 performed a reconnaissance as the sea was too rough for the seaplanes to be launched. No. 67 saw two submarines but not being equipped with W/T could not report until her return. This demonstrated the absolute necessity for W/T in reconnaissance machines.[22]

Leven is situated on the North bank of the Firth of Forth, North-East of Edinburgh. Blue Fleet had three seaplanes here including Short No. 42. No flying was carried out on the first day of the manoeuvres but on the 25th No. 42 flew over the Firth of Forth for 1 hour 34 minutes but sighted nothing.

There was plenty of aerial activity on the 26th. No. 67 did a scouting trip towards Cockle Gatway[23] and observed two submarines attacking *Maidstone.* No. 29 scouted towards the Wash and searched anchorages off the North coast of Norfolk without success in a flight lasting 3½ hours. No. 20 started a reconnaissance at 4 p.m. but was forced to alight at Winterton *where the machine, one Lieutenant and one Petty Officer were captured by Blue Territorials. Some difficulty was experienced in getting the machine back to the air-station. One seaman did exceptionally good work in swimming out (after the*

Above: Short S.56 with its serial No. 20 on the rudder. These early Shorts were clear doped with grey metal panels and varnished wooden coamings. Even the weather could not stop crowds from gathering wherever the RNAS seaplanes arrived.

dingy had swamped in the surf) in a strong tide at 10.30 p.m. with a rope to attach to the Seaplane.[24]

Hermes was also active again on the 26th but heavy seas prevented it launching a seaplane until late in the afternoon when Short Folder No. 81 was sent off. Difficulty was experienced in getting off the water and on alighting she bounced so heavily that the undercarriage was badly damaged and she had to be hoisted in for repairs. Owing to the damage done to the aircraft, comparative little flying was done at Great Yarmouth during the second phase of the program.

On No.81's final flight, leaving the ship at Yarmouth at 09:30 on 1 August, *a lovely day, making a good 75 miles an hour, at three to four thousand feet up; sometimes we had to come down to within 500 feet to get clear of the clouds and to get a good view.* Samson was to fly out for 50 miles then turn back. A brief undecipherable message was received shortly after the machine had turned back, then silence. The radio staff on *Hermes* were able to estimate a position and a destroyer, *Mermaid*, was sent to search for the missing machine with *Hermes* following as soon as she had raised steam. Fitzmaurice later reported that, *I had been sending signals by W/T all the time and had just reported a tramp on the starboard bow, when the engine suddenly stopped.*

We unfortunately made a bad landing and completely smashed our undercarriage and part of the lower plane and tail, fortunately, however, the tramp came to our rescue and hoisted us onto her poop.

She was a German timber boat, the Clara Menig, *bound from the White Sea to Tilbury; they were most kind and hospitable, but could not go out of their way to take us to Yarmouth.*

A few hours later the Hermes *hove in sight, having realized from our stoppage of signals that something was wrong; we hoisted out again from the* Clara Mennig *and hoisted into the* Hermes, *having had a very interesting day.*[25]

Hermes came up with the tramp, a German timber carrier *Clara Mennig*, a few hours later. *Hermes* recorded that she had sent No. 81 to scout East, and after a flight of about 50 miles No. 81 turned back. *Fourteen minutes after she turned,* Hermes *received an undecipherable message by wireless telegraphy and then nothing more.* It was concluded that the machine had come down. From this message *Hermes* worked out the approximate position of the machine and when it was found, it was within two miles of the projected position. *The cause of the accident to No. 81 was as follows: - A piece of tin in the inside of the bonnet got loose and broke four ignition wires, thus putting four cylinders out of action. The engine then stopped and a landing was made. In making the landing two float struts were broken and in hoisting the machine in and out of the steamer, &c., further damage of a minor nature occurred. The ship returned to Yarmouth and shortly afterwards the Manœuvres terminated.*[26]

Despite these experiences, aircraft came out of the manoeuvres with enhanced prestige. The weather had included fogs, driving rain, heavy seas and many valuable lessons were learned. *The experience gained during the Manoeuvres shows the necessity of fitting all Aircraft with Wireless Telegraphy.* Only the large seaplanes were capable of standing up to the rough conditions. The Maurice Farman floatplane at Cromarty air station had sprung floats and rode well at her moorings despite the bad weather. If No. 81 had had this fitting *it is probable that the chassis struts would not have carried away so frequently.* The folding wings of the

Right & Below: Short seaplane No. 20 come to grief at an airfield judging by the background. Machine is on a trolley but the undercarriage struts are broken. Sturtivant & Page record that the chassis was wrecked by 1 October 1915, however the machine had conducted patrols since the outbreak of the war and was overhauled and repaired on a number of occasions. No. 20 even reported a hostile submarine on 29 July 1917. The serial may be seen on the upper surface of the top wing tips.

Right: Although not clear, this photograph appears to show No. 20 landing on land. There appears to be some sort of undercarriage attached to the floats. Unfortunately, whether this photograph is related to the previous one is unknown.

Shorts appeared to work satisfactorily. *The folding wing type will certainly be the most suitable for ship work.*[27] Longmore *always regarded 1913 as the year when the Navy began to be interested* in what aircraft could do for the Fleet as a result of these manoeuvres.[28]

Although *Hermes'* floatplanes failed to make any significant contribution to the results of the 1913 manoeuvres, the extended period of operations was a watershed in the development of naval aviation. She continued operating her floatplanes into October, spending some days in Scapa Flow. About 20 further flights were made although only one, again by the Caudron on 3 September, used the platform. Methods of operation of floatplanes were established that, with minor changes, were to stand the test

Left: No. 20 taxying onto a beach.

Below & Facing Page, Above: No. 20 at Bridlington. It has the serial under each lower wing, that on the starboard side facing the other way from that on the port wing.

of war. In addition, the way ahead in naval aircraft design was clearly delineated. *Hermes* paid off into reserve at the end of the year.

At the end of the year Short No. 20 was one of the three aeroplanes at Great Yarmouth along with Nos. 29 and 69, a Maurice Farman seaplane and landplane respectively. *No. 20 occasionally flew, but No. 29 hardly ever. To start No. 20 was a feat of strength and was rarely accomplished in a short time.*[29]

Launching seaplanes from a beach required the use of a trolley. Trolleys either fouled the seaplane and or required many hands to manoeuvre them. A slipway would enable the seaplane to be launched in deeper water and generally smoother water, but still required the use of a trolley. Lt Cdr R Gregory, Eng Lt E.W. Riley (of *Hermes*) and a Mr White, a Chatham Dockyard official, proposed a wheel attachment to the floats such that it could be used as a land machine with the wheels being jettisonable when required to alight on water. A sprung disk wheel was fixed to the tail float. The device was known as the G.R.W. Wheel Float Attachment, after the initial letter of the three inventor's surnames. The Admiralty gave permission for the device to be trialled on Short No. 20.

The engine was started and the machine taxied along the concrete and turned down the slipway. She seemed to steer very well with the tail. Lieutenant Commander Gregory went down the slipway at a fair speed and straight on over about 10 yards

Left: No. 20 in full war regalia, with the RNAS red ring insignia on the upper surface of both wings, Union flag on the rudder under the large serial; at Great Yarmouth, August 1915.

of shingle into the water. Although travelling at a very fair speed (probably between 20 and 30 miles an hour) when leaving the slipway and striking soft sand the machine did not seem to have any tendency to get her nose down. Just before taking the water it appeared as if the machine were actually flying for about 2 feet. She took the water very easily and the wheels were then released as before. When Lieutenant Gregory returned, the machine was turned around so that her tail was on the beach, and there being little or no surf, the wheels were attached without difficulty. He then took the seaplane out to sea a little way and turned her around and charged the straight at the beach. Just at the water's edge the beach had formed a ledge and the wheels striking this caused the tail to lift uncomfortably high – at one moment it almost looked as if she would turn over on her nose – but it came down all right, and the pilot ran straight up the slipway to his shed under his own power.[30]

Further experiments were carried out with the device but the outbreak of the war saw the experiments being abandoned.

No. 20 carried out coastal patrols from Great Yarmouth after the outbreak of war. On 12 June 1915, the venerable Short No. 20 with crew of Flt Lt Chichester Smith and PO Jones, was sent out after a submarine had been reported off the coast. Two Henri Farman seaplanes also joined in the search but their efforts did not locate any submarine.

The general routine of patrols was a submarine patrol on the Short some time during the forenoon on those very rare occasions when the seaplane would get off. (Author's underlining) *It was a two-seater machine that would only carry the pilot, and was noted for its capacity to sling oil. The pilot flew with two 16-lb Hale bombs on his lap, bomb frames being too scarce in those days to fit an ordinary seaplane, besides the extra weight stopped*

Serials		
Serial	**Contract**	**Notes**
10		S.41. Delivered by 27.03.12. Probably destroyed 04.11.15.
20	Contract P43037/13	S.56. Delivered 21.04.13. Deleted 14.02.16.
21		S.57. Delivered 22.04.13. Deleted 07.02.15.

her "getting off". On one occasion it was known to have reached the prodigious altitude of 1,000 feet; the pilot, a very harmless, innocent "quirk", hardly fledged, straight from Chingford, having struggled with the machine for two hours, came home so full of life that he told everyone, even the Commanding Officer, of his experience. "What the hell do you mean by doing a submarine patrol at that height for? You would never see a submarine from 1,000 feet, even if it was there!" and that was all the thanks he got from his C.O.[31]

No. 21 was delivered to Grain in April 1913. By February 1914 is was *old and soggy* and returned to the makers for reconstruction.[32] It was fitted with a land chassis by July 1914. Shortly before midnight on the night of 14/15 August 1914, a report of an airship sighting near Gorleston-on-Sea steering south led to No. 21 being readied with two parachute flares loaded with lead shot. The weather conditions were ideal and it was thought that the machine would have no difficulty in manoeuvring to height if an airship was sighted. The report was false and the aircraft did not take-off.[33] During the daylight raid of 25 December, by a Friedrichshafen FF 29 floatplane, No.21 and Short Folder No. 121 that were stood by at Grain were put on emergency standby, but were not called on.[34]

Damaged on 8 September, when the undercarriage gave way, No. 21 was dismantled. Flt Lt S Pickles flew the repaired aircraft on 25 December 1914.Used as a Home Defence aircraft at Grain. In January 1915, it was apparently used in bomb dropping experiments. It was written off on 7 February 1915, due to wear and tear.[35]

When war broke out the RNAS had the following 17 Short seaplanes out of 31 seaplanes on strength:

6 Short 100-hp Gnôme engine.
9 Short 160-hp Gnôme engine.
1 Short Canton-Unné engine.
1 Short pusher gun-bus 200-hp Canton-Unné engine.[36]

According to C.F. Snowden Gamble, just before the war, a conference of officers commanding seaplane stations agreed that standardization of a type of scouting seaplane for war purposes was desirable. They were of the opinion that the 100-hp Gnôme Short was the most reliable machine then in service, but it should have sprung floats and folding wings.

End Notes:

1. History of Naval Aeroplanes and Seaplanes" by Colonel C.R. Samson, stamped 30/12/18. TNA AIR1/724/76/2.
2. "Early British Naval Aviation." This undated paper was written by the then Air Commodore Samson after his retirement. Copy in IWM via G Bruce.
3. Short, O. "Twenty-One Years' Progress of Seaplane Design." *Flight*, 3 January 1930, P.68.
4. *Catalogue of Naval and Military Aeroplanes*, Short Brothers, Naval and Military Aeroplane and Dirigible Balloon Constructors. Copy in Sir Austin Robinson's family papers.
5. "Early British Naval Aviation" *Op Cit*.
6. *Ibid*. In his Log Book Samson wrote *10 AM. Took H 1 (100 HP Gnome Tractor) off water in Portsmouth Harbour, flew around fleet in Harbour 3' in 4 mins Landed on water. 1st Time I have been off water & landed on water*. Samson's underlining. Samson's Log Book in IWM.
7. E-mail from G Bruce. 13.07.2020.
8. Quoted in King, H.F. *Armament of British Aircraft 1909–1939*, Putnam, UK, 1971. P.307.
9. Of No. 10 the late J.M. Bruce wrote that he never *ceased to be amazed at the apparently endless changes of shape that Short No. 10 undergoes – they must have modified it roughly every other month!* J.M. Bruce to G.S. Leslie 22 .09.1997. *The changes from the earliest No. 10 are so extensive that it could hardly be the same airframe throughout its career!* J.M. Bruce to G.S. Leslie 2.11.1997.
10. Despite Samsons assertions that this was the first firing of a machine gun from an aeroplane, Capt Chandler of the US Army fired a Lewis Gun from a Wright Model B at College Park on 7 June 1912. See https://www.thisdayinaviation.com/tag/charles-deforest-chandler/
11. King, H.F. *Op Cit*. P.307.
12. Samson's log records further Maxim firings from the S41: 27 February 1913: Susan fired 53 rounds of Maxim. 12 March 1913: further firing with Susan as gunner. 5 May 1913: Lt (G) Clark Hall as gunner who *fired it at sea, one belt and a quarter*. Notes from G Bruce, July 2020.
13. Notes from Gordon Bruce from TNA AIR1/3349/15/227/4. The Maximum used by Susans and Clark Hall from S41 was an 'antique.' The Maxim's machine gun was recoil operated and belt fed. Vickers built it to take the standard Army's 0.45 inch caliber, same as the Army's standard infantry rifle ammunition. In 1885 the Army adopted the 0.303 inch rifle and Vickers changed their machine guns to take the new ammunition. Early guns were modified from 0.45 to 0.303 caliber. In the early 1900s, Vickers developed a light-weight version of the gun.
14. Quoted by H Woodman in 'Armament Notes.' Typescript in RAF Museum JM Bruce Collection Box 20.
15. In their issue of 21 March 1914, *Flight* announced that

Specifications Short S.41

Source	1		1		1		2
Engine	80-hp Gnôme		100-hp Gnôme		160-hp Gnôme		100-hp Gnôme
Span upper	40 ft 0 in	13.72 m	50 ft 0 in	15.25 m	50 ft 0 in	15.25 m	50 ft
Span lower							34 ft 9 in
Length	35 ft 0 in	10.67 m	39 ft 0 in	11.0 m	45 ft 0 in	13.72 m	39 ft
Height	10 ft 0 in	3.06 m	11 ft 9 in	3.58 m	11 ft 9 in	3.58 m	11 ft 9 in
Chord							6 ft 11 in
Gap							7 ft
Unloaded Wt.	1,200 lb	545 kg	1,700 lb	764 kg	2,000 lb	909 kg	
Empty Weight							1,100 lb
Max approx	65 mph	105 km	60 mph	97 km	74 mph	120 km	60 mph
Min approx	50 mph	80 km	50 mph	80 km	56 mph	90 km	
Duration	4 hrs		5 hrs		6 hrs		5 hrs
No of Crew	2		2		4		

Source:
1. *Catalogue of Naval and Military Aeroplanes*, Short Brothers, Naval and Military Aeroplane and Dirigible Balloon Constructors. Copy in Sir Austin Robinson's family papers. The same specifications are given in *Jane's All the World's Aircraft 1913.*
2. J.M. Bruce data. This data corresponds with that in the Short catalogue drawing that shows what appears to be Short S 41 No. 20.

Short Brothers *have developed a method of folding the wings of waterplanes, by means of which the overall width is reduced to about 10 feet.* The article noted that it was not at liberty to discuss this feature before now, the advantages of which, *especially for machines which have to be housed on a ship, are obvious.* "Seaplanes with Folding Wings." *Flight*, 21 March 1914. P.313.

16. "Naval and Military Aeronautics," *The Aeroplane*, 19 March 1914. P. 336.
17. Gorden Bell had become Short Brothers first test pilot on staff after Frank McClean found the task too onerous for him to continue to do it on a voluntary basis. At this time Bell was only part-time and still a freelanced test pilot for other manufacturers.
18. Snowden Gamble, C.F. *The Story of a North Sea Air Station*. Oxford-London University Press. UK. 1928. P.44.
19. TNA AIR 1/626/17/14.
20. Naval Manoeuvres July-August 1913. Appendix No.1. Report of Captain Hermes on use of aircraft during Manoeuvres. TNA AIR1/626/11/46, via G Bruce.
21. Although he did fly it on 21 July – *1 hour in "Borel" hydro trying to get off water. Engine 1120 revs. Just got off twice, would not fly.* Samson's Log Book in IWM.
22. Naval Manoeuvres July–August 1913. Appendix No.1. Report of Captain of Hermes on use of aircraft during Manoeuvres. TNA AIR1/626/11/46, via G Bruce.
23. A Gat was a channel. [Cockle Gatway was the principal entrance to Yarmouth Roads.]
24. "Report of the Captain of *Hermes* on Use of Aircraft during Manoeuvres." *Op Cit*.
25. Quoted in Gamble, *The Story of a North Sea Air Station*, P.55.
26. Naval Manoeuvres July–August 1913. Appendix No.1. Report of Captain of Hermes on use of aircraft during Manoeuvres. TNA AIR1/626/11/46, via G Bruce.
27. Goodall, M.H. 'Early Days, HMS Hermes and the 1913 Naval Manoeuvres', *Cross & Cockade Great Britain Journal*, Vol. 7 No. 1, 1976. P.154, contains the original reports of the manoeuvres. Also, Snowden Gamble, C.F. *Op Cit.*
28. Longmore, A. *From Sea to Sky*, Geoffrey Bles, UK, 1946. P.29.
29. Snowden Gamble, CF. *Op Cit.* P.64.
30. Report by sub-Lt Kershaw. Quoted in Snowden Gamble, C.F. *Op Cit*. P.69.
31. Snowden Gamble, C.F. *Op Cit.* P.133.
32. Sturtivant & Page. *Op Cit*. P.23.
33. J.M. Bruce notes from TNA AIR1/186/15/226/3. RAFM J.M. Bruce Collection Box 4.
34. J.M. Bruce notes from TNA AIR1/186/15/226/3. RAFM J.M. Bruce Collection Box 4.
35. Also, Cole, C & Chessman, E.F. *The Air Defence of Britain 1914–1918*, Putnam, UK. 1984. P.20.
36. Sturtivant & Page, P.23.
37. Snowden Gamble, C.F. *Op Cit*. P.82.

Chapter 4: Short Admiralty Type 74 Tractor Seaplane

The story of these Short seaplanes, the first to give its number to a class of aeroplanes, illustrates the problem with the different Short and Admiralty designations. These four seaplanes were ordered as Admiralty 74 Type, however, the first Short Folders, were also given the same designation, and are covered in Chapter 5.

S.69 to S.72 – Admiralty 74 Type Tractor Biplane Seaplanes. RNAS Nos. 74 to 77

Four Short Admiralty 74 Type Tractor Biplane Seaplanes, powered by 100-hp Gnôme Monosoupape rotary engines, were ordered by the Admiralty under Contract CP38479/13, and received the RNAS serial Nos. 74 to 77.

Bearing a resemblance to No. 42, the Type 74 had three-bay non-folding wings. Typical Shorts construction was followed with a box girder fuselage, ply panels around the cockpits, metal engine cowls for the 100-hp Gnôme ten-cylinder rotary engine. The rear fuselage was fabric covered. Single-acting ailerons were fitted to the upper wing. A triangular fin was fitted in front of the large rudder. No. 74 was first flown by Gordon Bell on 4 January 1914, with the remainder of the contract following through the next four weeks.

The four seaplanes attended the 1914 Fleet manoeuvres and the Royal Review Flypast. No. 74 served at Grain and Westgate and was finally written off after an aileron jammed on 29 March 1915. Grain recorded No. 74 on a submarine search on 10 August 1914. No. 75 patrolled with Short No. 19 and Short No. 905, the Nile pusher biplane, on the next day. On 22 August 1914, Nos. 74 and 77 *were prepared with spare engine and necessary stores for transhipment on arrival, to SS* Indrani. When the ship arrived at 1900 it was found that to stow the machines onboard with any safety Nos. 74 and 77 would have to be dismantled. A carpenter party worked all night dismantling the machines and the stores were sent on board by steam boat lighters. The ship left the next day for Yarmouth carrying the two machines, spares and party. That same day the SS *Empress* arrived Sheerness at 0630 from Ostend with Nos. 75 and 76 on board, apparently returning them for repair.[1]

No. 76 had been taken on charge at Grain in February 1914, and later went to Dundee in September. In early December it went for a major overhaul by Blackburns, returning to Dundee on 30 January 1915. Three weeks later it was seriously damaged, and was probably never repaired. No. 77 flew North Sea patrols from Dundee. Flt Lt Henry Doone Vernon, with Flt Lt Basil Drummond, never returned from their patrol on 29 September 1914, and are presumed to have been forced to alighted and were drowned. They have no known grave. None of the Type 74 seaplanes survived past March 1915.

Barnes shows a photograph of Samson taking the First Lord of the Admiralty, Winston Churchill, for a flight in No. 76. Samson's Logbook shows that he did not fly between the 4th to 15th June. He did take Churchill for flights on 16 May and 4 June 1914, but both of these flights were in landplanes.[2] Lt Jack Seddon recalled a most exciting flight with Churchill when he, and G.V. Fowler, each in his own seaplane, *Short 100's*[3], flew Churchill and his secretary, Edward Marsh, from Gravesend to Grain. The flight was made against a strong south-easterly wind in rapidly deteriorating weather. After alighting safely after a flight of about 45 minutes, Seddon's aircraft, beaching trolley included, was blown over the sea wall as it was being brought up the slipway and more or less wrecked. Fowler took nearly three hours for the 16 miles from Gravesend as he had chosen to fly high to avoid the bumps. Seddon flew low and it was so rough he recorded that if he was his own passenger, he would have been sick, however Churchill was a much better sailor than he was and never mentioned it.[4]

After his successful torpedo drop from Short No. 121 on 28 July, 1914, Lt Arthur Longmore recorded that during the forenoon following machines left for Grain - Short Nos. 74, 75, 76, and 77, Henry Farman F.22H Nos 141, 142, 143 and No. 48 Borel monoplane seaplane. By evening No. 76 was at Littlehampton with engine trouble, No. 74 at Dover with float problems, Nos. 75 and 143 at Dover with engine trouble, No. 142 at Dover, No. 141 was wrecked at Yarmouth, and only No. 77 arrived at Grain.[5] This exercise is a typical commentary of the problems with reliability that were present at the opening stages of the war.

S.128- Admiralty 74 Type Tractor Biplane Seaplanes. RNAS No. 183

Built under Contract CP46634/14, this lone Short 74 Type was originally allocated to Dundee but was delivered to Grain instead on 24 November 1914. By January 1915, it was being used for instruction, then in June it was sent to Gibraltar. What its purpose there is unknown. It was being overhauled *(Taking down)* from 16 to 20 July, on the latter date its fuselage was being trued up.[6] On 10 August it was reported as *Erection completed. Engine fitted and tested – Satisfactory*. The following day a 10 minute flight was recorded: - *Flt.Lt. England and pass First flight since overhaul.*

Right: Two Short Type floatplanes in shallow water. Serial No. 74 to the right. Note how ailerons reflect sunlight.

Left: Boy scouts assist in pushing Short Type 74 floatplane serial No. 74 across the sand on it beaching trolley. The way both ailerons hang down, due to their being unconnected, was a characteristic of early Shorts.

Right: Serial No. 74 on Deal Beach.

Left: Serial No. 75 non-folding 100-hp seaplane at Spithead in July 1914.

Machine vibrating badly.[7] It was still recorded at Gibraltar on 30 November 1915, but not by March 1916. Its fate is unknown.

When war broke out the RNAS had the following 17 Short seaplanes out of 31 seaplanes on strength:

6 Short 100-hp Gnôme engine.
9 Short 160-hp Gnôme engine.
1 Short Canton-Unné engine.
1 Short pusher gun-bus 200-hp Canton-Unné engine.[8]

Westgate Naval Air Station

W/Cdr G.R. Bromet recollected that by late 1915, Westgate had expanded enormously and included an aeroplane station as well as a seaplane one. Seaplane sheds and slipways had been built close to the original 1914 site and an aerodrome had been acquired about a quarter of a mile away towards Margate. At the aerodrome we had an assortment of machines including B.E.2c's, Avro and Curtiss, and at the seaplane base we had three Shorts (150 Sunbeam) and one "Schneider Cup" Sopwith. As far as I can recollect we had no definite scheme of work. Our chief functions were to provide seaplanes for Channel and Thames patrols, and to act as guardian angel for Margate and other coast towns by having aeroplanes available for the pursuit of any hostile aircraft which appeared.

The armament of aircraft at this time was still very crude but important developments were just beginning to materialize. The principal offensive weapons fitted to both seaplanes and aeroplanes were the bomb (16-lb., 20-lb., and 65-lb.) and the grenade. During my time at Westgate Lewis guns were introduced and we had them fitted to the B.E.2c's, but all the other machines had to be content with grenades, darts, bombs or sporting guns.

Source: War Experiences of W/Cdr G.R. Bromet 1914–1918. (2nd Course) TNA AIR1/2387/228/11/30.

Endnotes

1. Notes from TNA AIR1/ 186/15/226/3 by J.M. Bruce. RAF Museum J.M. Bruce Collection Box 4.
2. Notes from Gordon Bruce. Samson's Logbooks are in the IWM, London.
3. Sturtivant & Page suggest that the flight with Churchill from Tilbury to Grain on 7 February 1914, may have been in No. 78. However, this was a 160-hp Gnôme powered machine. To confuse the matter more, they also have Samson recording the machine as having a 100-hp engine!
4. De la Fereté, P.J. 'Churchill the Airman', in Eade, C (Ed). *Churchill by his contemporaries*, The Reprint Society, UK, 1955. P.103. Note that Seddon's surname is incorrectly spelt as Sedden.
5. A Longmore's Log Book, National Maritime Museum, Greenwich, via G Bruce.
6. 'Diary of Events from 15 to 21-7-1915.' TNA AIR1/665/17/122/717.
7. The 'state of Aircraft on Wednesday 11/8/15.' Gives No. 183 as Ready for service with Gnôme engine No. 4104 installed. 'Diary of Events W/E 11-8-1915.' TNA AIR1/665/17/122/718.
8. Snowden Gamble, C.F. *Op Cit*. P.82. J.M. Bruce was skeptical of this listing. See his letter in reply to the article 'The Short Story' in *Air Pictorial*, March and April 1957.

Right: Serial No. 76 appears to be engineless Note the aerial wheel on the side of the cockpit. The machine behind has retained its engine.

Below: Working on a Type 74 seaplane at the same location. The ailerons are not interconnected.

Right: Serial No. 76 floating off a beach. Note the Shorts' logo on the bottom of the rudder.

Left: Serial No. 77 on a beach with the engine requiring attention. As appears usual, Boy Scouts have been attracted to the spot.

Right: Serial No. 76 on a sandy beach with the sea appearing to be getting a swell up. The engine is exposed with the top cowl panel removed.

Left: Serial No. 77 on the shore at Port Victoria, Isle of Grain, with the engine running.

Right: The pilot of this Short Type 74 is believed to be F.W. Bowhill. Note the weighted aerial wire at the tube attached to the lower longeron.

Below: A Short Type 74 seaplane taxying towards a wide wooden trolley. Once the machine had reached the trolley, it would be hauled up the slipway.

Chapter 5: The Short Folder Seaplanes

In July 1914, S/Lt John Tremayne Babbington was serving at the Isle of Grain when he received sealed orders from the Admiralty. He left Grain with two Short 160-hp seaplanes in order to establish a seaplane base at Westgate-on-sea.

The duties of the seaplane flight were simple: we were required to maintain patrols from Westgate to the North Hinder Lightship during the transit of the Expeditionary Force across the Channel to France. *Orders were received from Sheerness Air Office and, or direct from the Air Department, Admiralty.*

I do not recollect that the patrols were directly instrumental in assisting the Expeditionary Force, though they were maintained with great regularity in spite of the low powered Gnome-engined seaplanes with which the flight was equipped.

The armament of machines consisted of shot guns, rifles with rifle-grenades, revolvers, grapnels and improvised bombs – (generally sulphuric acid). The bombs and grapnels were destined for use against airships, though the poor performance of the seaplanes would probably have prevented their successful use.

There was no "slipway" at Westgate, though the sandy shore facilitated the launching and beaching of seaplanes except at periods of high tide. In general all facilities had to be improvised, including civilian guards, civilian crews to assist in getting machines in and out, etc.

Looking back, the outstanding feature of this episode – apart from the patrols and reconnaissance caried out with machines and material of undeveloped types – was the way in which our highly skilled mechanics and carpenters contrived to keep the seaplanes going under what would now be considered s impossible conditions. It should be added that during this short period one was conscious of fulfilling a definite purpose with a definite aim. – At later periods there were times during the course of the war when this consciousness of a definite functional use and responsibility were not always present.[1]

The Short seaplanes Babbington mentions were the Short Admiralty 81 Type Folders Nos. 119 and 120, that served at Westgate in August 1914.[2]

C.G. Grey, the editor of *The Aeroplane,* reported in the issue of 19 March 1914, on 'Seaplanes at Eastchurch.' This was the first occasion that the aviation press was allowed to report on the Short method of folding wings, *the British and German patents being now duly signed, sealed and delivered.*

On Wednesday of last week, the 11th, two of us visited Leys*down to see the flights of one of the latest Short seaplanes, 160-h.p. Gnome, which was just delivered to the Naval Air Service. From the Short Bros.' works at Eastchurch we were driven over to Leysdown by Mr. Oswald Short, where we found the machine on the beach ready to start. The regular process is for the seaplane to be assembled in the works. It is then put on a two-wheeled trolley, the tail is attached to a car, and it is towed to Leysdown, whence it is flown to the Isle of Grain. Sometimes the seaplanes are first tested with a land chassis, but in this case the machine was built straight onto her floats, and when we saw her on the beach she had never been in the air. Mr. Gordon Bell climbed in, with Mr. C.R. Fairey as passenger, the machine was pushed into the sea, Mr. Short and one of his men going up to their waists in the ice-cold water to clear the land-wheels.*

These machines have internal starting handles, and as soon as this one was well afloat, the engine was opened out and away she went down wind and diagonally across the run of the sea. Then, with a strong wind well on the port quarter, Mr. Bell picked her off the water and she lifted like a bird, after a shorter run than one usually sees when a machine starts down wind on a smooth aerodrome. After one or two curves to get the feel of the machine, Mr. Bell did a sharp right turn, coming back close along the shore, and after another circuit over the sea, made a couple of

Above & Right: One of the first Short Folders, Serial No. 81 or 82, at Eastchurch, 1913.

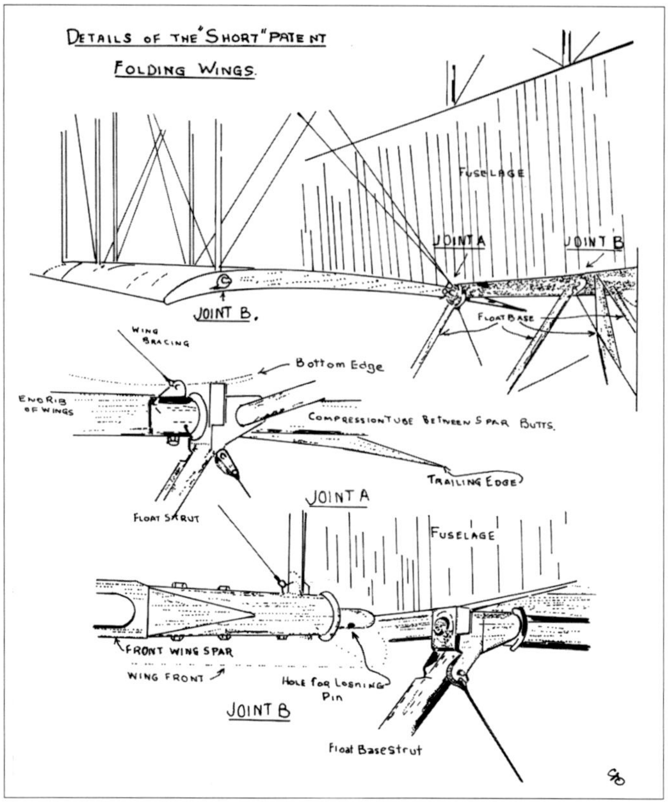

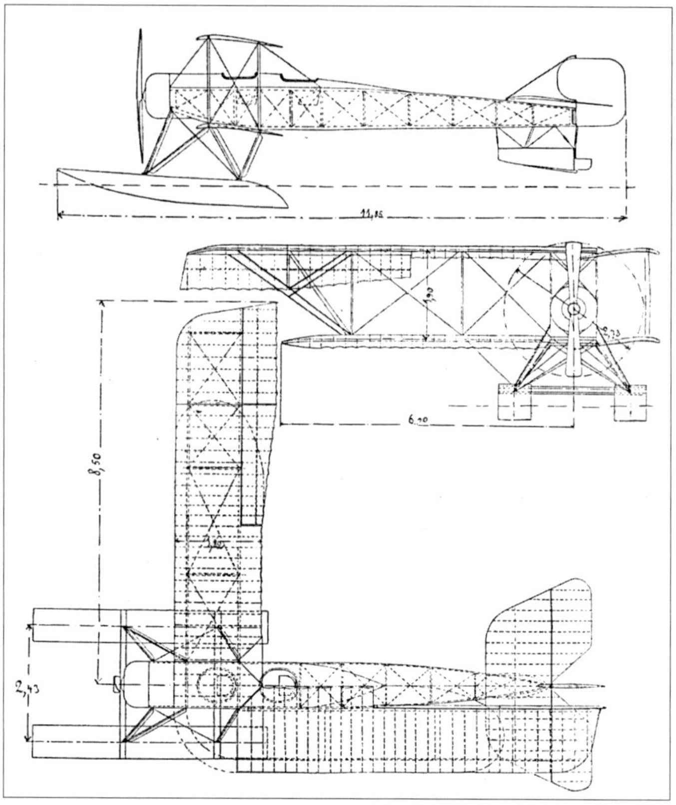

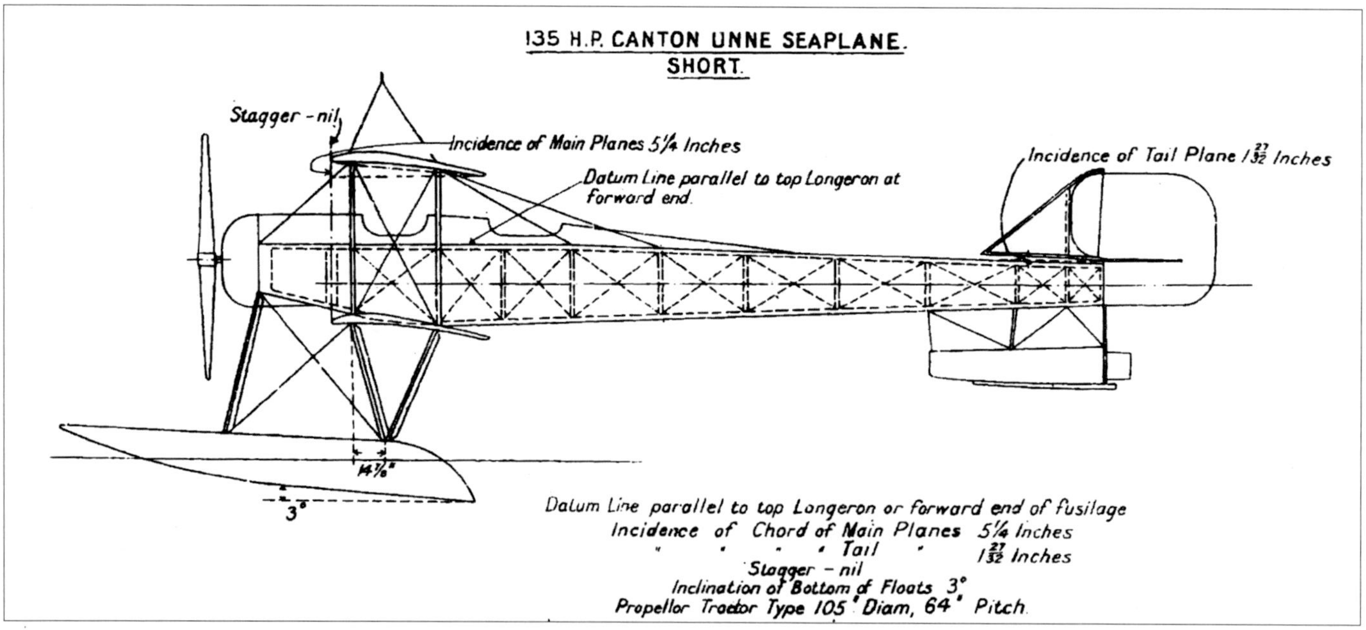

Above Left: The Short Folding wing mechanism. Redrawn from *The Aeroplane*, issue of 19 March 1914.
Above Right: This drawing of a Short Folder accompanied *The Aeroplane* article on the New Short Seaplane. It was reproduced in German magazines, such as *Flugsport*, as the *Wasserdoppeldecker von Short* with the Imperial dimensions converted to metric.
Above: The rigging instructions for the Short Folders were very basic. This drawing was reproduced for the 200-hp Canton-Unné model with only the notes changed.

sharp spirals over the land, but within gliding distance of the sea, apparently to test her steadiness in land gusts. Then he went off to Isle of Grain.

Grey commented that despite the weight and size of the seaplane, it handled as well as the best land machines *although it was her first trip and she was untuned.*

Right: Short Type 64 Folder, Serial No. 81, operating with HMS *Hermes*, July 1913. The machine has the typical unconnected ailerons and the Shorts' logo on the rudder but no serial number. The pilot in the rear cockpit is Lt C.R. Samson. Note the fretted cowling.

Left: Serial No. 81 being hauled aboard HMS *Hermes*, 1913.

Remembering Grey's attitude that British was best, his comments are still worth recording. When Gordon Bell returned from Grain, he *told us that he had passed her through the tests with ease. He had not the official speed, but estimated it at about 77 to 78 m.p.h., which I should say is about right. … It seems that the Short Brothers have good right to claim that they make the most efficient seaplanes in the world. At any rate, if there is anything to beat it one would like to know of it.* A number of photographs of a two-bay machine with wings in the open and closed positions accompanied the articles.[3]

Under the title 'The New "Short" Seaplane' a detailed description of the new machine was presented by W.H. Sayers. The seaplane was powered by a 160-hp Gnôme engine that had a large streamline funnel over the engine.[4] *It has a very wide and relatively shallow fuselage, rectangular in section, built with ash longerons and spruce struts. U-bolts are employed for fastening the fuselage struts in place and to carry the box wiring. The 14-cylinder Gnôme is mounted in steel housings within the fuselage members. The motor is surrounded by a sheet metal cowl, in the top of which air holes are cut for cooling, and on the top is fitted the chimney already referred to to carry away the both air and exhaust gases. The rounded engine cowl is carried back and tapered off into the fuselage proper behind the pilot's seat, being cut out and bellied up to serve as wind-screens at the passenger's and pilot's seat.*

At the rear of the fuselage, is the empennage, consisting of a large cambered fixed tail-plane, followed by divided elevator flaps, a large fixed vertical fin, and a balanced rudder of generous

Above: Hauling No.81 aboard HMS *Hermes*. The starboard wing is not folded and there is a broken strut to the rear of the starboard float. This may be the incident of 27 July 1917, when the machine broke struts. Samson, in his white helmet, stands on the port float.

Left: Samson in No. 81 alongside HMS *Hermes*. The wheel control is well shown in this view. The starboard wing tip appears to have suffered damage.

Above & Left: The wings are folded on this two-bay Folder before attempting to hoist it aboard. The late J.M. Bruce considered that the plain shiny single-curvature top cowling may have been a replacement knocked up on board *Hermes* after the accident of 22 July 1913, when a gale blew down the canvas shelter housing No. 81.

Left: No. 81 under tow with a man on the float keeping an eye on proceedings. The number of sailing ships in the background is an indication that the days of sail were not over.

Left: HMS *Hermes* with a Short Folder on the aft deck. The machine displays no serial.

size. Built into the fuselage, behind the engine, is the centre cellule to which the main wing structure is attached. This consists of the ordinary short top section of wing, mounted on four steel struts; but the bottom rectangle of the structure, instead of being within and part of the fuselage, is beneath it, and consists of a pair of heavy steel tubes in line with the front and rear spars of the bottom planes, joined near the sides of the fuselage by a couple of lighter steel tubes. The corners of this rectangle of steel tube are immediately beneath the struts from the top plane section, and are joined to them with substantial steel clips.

The reason for this detailed description was that this was the first time that the Short patented folding wing system had been allowed to be described in the aviation press. Sketches accompanied the article showing the method that Shorts had devised to enable the wings to fold.

The cross-tube that corresponded with the rear spars was fitted with a 'snout' of steel. This 'snout' fitted into the end of a long steel box fitted over the end of the bottom rear spar of an outer wing section, and was united by means of a large steel eyebolt. The face of the steel box was cut away and

Above: No. 119 taxiing. The 'chimney' exhaust has a single white band. Each of the four Folders taking part in the Royal Naval Review of 18–22 July 1914, had rings, one to four, painted to identify the individual aircraft, similar to the rings painted on the funnels of RN ships.

Above: Launching No. 120 at Westgate.

Right: No. 120 moored out at Westgate. Note the two white bands around the exhaust chimney required by the 160-hp Gnome rotary engine.

Left: The crew and their helpers in the rowing boat appear to be having trouble in handling No. 121. The man on the wing is carefully walking on the wing spar.

the joint became a hinge. The upper rear spar was similarly treated, the spar of the centre section carrying the "snout."

The front spars of the outer wing sections were also fitted with long steel boxes that terminated with steel pins. With the wings in the flying position these pins enter the front cross tube and are held by a heavy locking pin (about ½ in dia) that passed through both members of the joint. Each wing was fitted with a heavy drift wire to prevent it folding back. *Thus by pulling out four pins and undoing four wires, the wings may be folded back parallel to the fuselage with the trailing edges of the bottom planes just not meeting under the centre line of the fuselage, in a few seconds.* All four locking pins could be taken out or replaced from the passenger's seat. A winding reel was also fitted here with which the wings could be wound in or out, and a simple catch fitted at the rear of the fuselage clipped the end rear interplane strut near the spar, holding the wings securely in place when folded. The whole of the interplane bracing was not affected so there was no chance of the truth of the planes being interfered with. *Though not easy to describe clearly, the whole arrangement is simple and very thoroughly carried out.*

The wing leading edges were of spruce with wire trailing edges. The interplane struts were oval steel tubes. All lift wires were multi-strand cable doubled. Large unbalanced ailerons were fitted to the upper wing that had a large overhang. *The Aeroplane* thought that these would *have proved inadequate, or at least heavy to work, in a machine of such speed and weight, but Mr. Gordon Bell, whose partialities are rather towards a lively machine, says he knows none so easy to handle, no mean testimony to the excellent balance and high aerodynamic qualities of these seaplanes.*

The unstopped, sprung wooden floats had their bottoms reinforced with metal. All the struts were mounted on universal joints at their tops, those raking forwards being of spruce, the others being steel tubes with streamline wooded fairings. The floats were divided into watertight compartments, each having a drain tube, all of which were brought into a single valve-box from where the whole float could be blown dry. The box was fitted with a pressure release valve to reduce the effect of the reduced atmospheric pressure at high altitudes. The floats were placed well forward, *and a fair-sized tail-float is fitted to support the tail when at rest.*

The principal dimensions given are: - Span top 56 ft; Span lower 40 ft; Chord 6 ft; Length 39 ft; Area 550 ft^2; Weight empty 2,000 lbs; Useful load Passenger & Pilot 3320 lbs, Fuel & Oil 600 lbs; Wireless 120 lbs – total 1,040 lbs. 160-hp Gnôme engine. Capacities: Fuel 65 gals – 5 hours flight; oil 15 galls; speed 78 mph, climbing speed 600 ft/min to 1,000 ft.[5] A set of *authentic scale drawings of the new Short Seaplane with folding wings* and dimensions accompanied the article, but were not of the machine described but of one to be powered by a 200-hp radial engine.[6]

These two articles show some of the problems in trying to determine what Short machine is being described. The photographs in *The Aeroplane* show a two-bay machine. Grey noted that the machine illustrated was a similar type but with a 100-hp engine. We have two experts describing two different machines, and the scale drawings show another machine entirely. It is concluded that the detailed description appears to have been taken from the 100-hp machine and applied to the 160-hp machine as far as the folding wing mechanism. The delivery date of 11 March 1914, does not match any dates recorded by Sturtivant & Page.[7]

The Short folding wings mechanism was developed by Horace Short in 1912. The wings were hinged on the back spar and folded backwards. Bayonet joints on the front spars locked the wings into position for flying and the bracing had quick-release catches. This device was patented but, according to J.M. Bruce, the only payment that Short received was £15 that T.O.M. Sopwith paid for use of the patent, presumable for his type 807 and 860 seaplanes.

Samson wrote that *Undoubtably the next big step in Aviation was the folding wing invention of Short Brothers; this provided just what the navy wanted, as with it we could carry seaplanes aboard ships. I tested the first one in the naval Manoeuvres of 1913. This was the best seaplane produced to this date. She flew splendidly, got off well and was well ahead of her time in every particular.*[8]

Folding Wing Aeroplanes

The Short seaplanes were not the first aeroplanes to have folding wings. The De Marcay-Moonen monoplane, described as the Malicet-Mooney monoplane, was featured on the front cover of *The Aeroplane* for 7 December 1911, and that of *Flight* for 20 January 1912. This monoplane with a span of 13.5 metres underwent trials in late 1911 at Issy. It was stated to be a collaboration between Baron Edmond de Marcay and Emile Moonen. The main planes swung on two masts that were set at an angle through the mainplanes. The folding of the wings was done from the pilot's seat. Despite this, all three Short Brothers were granted patents Nos. 1,792, 15,727 and 28,610 in 1913, covering their wing folding mechanism. These and other contemporary aeronautical patents are detailed in Harold Penrose's Volume 1 of his detailed history of British aviation (*British Aviation – The Pioneer Years,* Putnam, UK, 1967).[9]

S.64 & S.65 – The First Folders RNAS Nos. 81 & 82

As described in Chapter 1, under the 'Report of the Committee of Imperial Defence Paper 139-B', the Navy proposed to acquire various aircraft in order to ascertain the best types for Naval services. In the Shorts' Order Book Job No. 337 (S.552) is only described as *Admiralty.* However, Job No. 796 was for two *160 hp Tractor Hydro Aeroplane. Navy.* This was noted as *These two Machines are to be made in place of One Twin Engine Machine No. 552 Job 337.*[10] The two new replacement aircraft were S.64 and S.65. The reason for this change is found in the official report *Naval and Military Aeronautics – 1913,* wherein it is noted that *Three twin-engined machines had also been flown but the results were not as satisfactory as was anticipated.* [11] These three were all Shorts; the S.39 Triple Twin, the S.47 Triple Tractor and the S.46 'Dirty Double'.

Although the Short twin-engine machines had not proved successful, progress had been made during 1912 in single-engine tractor floatplanes. The fact that one twin-engine machine was equivalent to two single-engine machines points to the availability and cost of engines at this time.

Horace Short soon built a machine that utilised his wing folding device. This was No. 81 (S.64), a floatplane that was accepted at Eastchurch on 17 July 1913; and was closely followed by No. 82 (S.65) in October. Both utilised the 160-hp Gnôme twin-row fourteen-cylinder rotary engine. While similar to the Short Admiralty 74 Type, particularly in the design of the tail unit and fuselage, these seaplanes had high aspect ratio wings arranged in two-bay rather than the three-bay of the 74 Type.[12] The biplane had steel-tube interplane struts and strut braced extensions. Stronger and larger rubber-sprung floats were fitted, and the balanced rudder had a triangular fin in front. The pilot occupied the rear cockpit. In common with other Shorts, the wide ailerons were uncompensated and hung down when at rest.

These aircraft, although unimpressive in appearance, were important historically as it introduced the RN to folding wings on its aircraft. According to Barnes, these were simple folding wings that had to be secured by personnel outside the seaplane

The Royal Navy Manoeuvres of 1913

The Aeroplane reported on 24 July 1913, that *Early last week H.M.S. "Hermes" returned from Yarmouth where she delivered*

Above: Short Folder S.84, Serial No. 121 at the Isle of Grain, 1915. This was the first British seaplane to launch a torpedo. Note the arched spreader bars between the floats. (W.G. Moore Photograph)

Right: The arched spreader bars for the floats of No. 122 are well displayed in this view.

Left: The remains of No. 813 or 817 as recovered in the Netherlands. Both these machines went missing in action on 16 February 1916.

the two Caudron "amphibians." After coaling at Sheerness she took on board on Thursday a Borel sea-monoplane (80-h.p. Gnome) and a Short sea-biplane (160-h.p. Gnome), which is to be flown by Commander Samson, R.N., and left on the 19th for the North. This Short and one of 100 h.p. passed their acceptance tests on Thursday, the 17th. [13] This was Short No. 81 that was to be flown by Samson in the Royal Navy Manoeuvres of 1913, the first in which aircraft took part with the Fleet at sea. No 81 performed well in the manoeuvres and this is covered in Chapter 3.

No. 81 was seriously damaged on 4 September 1913, while being flown by Flt Lt F.W. Bowhill with a mechanic as passenger. No. 81 left the *Hermes* and *made a circuit over two battleships in the bay. When about 1,000 ft. up the rudder jammed and the machine made a steep dive into the sea, doing considerable damage. The pilot and passenger were picked up by motor-launches unhurt.*[14] It was repaired by Shorts and was at Grain in April 1914. Wrecked the following month it was deleted on 25 June 1914.

No.82, the sister seaplane, was tested by Gordon Bell on 2 October 1913. It participated in the Royal Review of July 1914. No. 82 was taken out by Gordon Bell at Calshot on Monday, 6 July 1914, having been delivered to Calshot by road the previous month. Arthur Longmore, then the CO of Calshot, wrote that *This machine is by far the best seaplanes I have seen yet at Calshot, she got off side to the wind easily. The mounting of her floats appears excellent.*[15] On the 22nd it was towed into Dover after an engine failure in the Downs. In August it returned to the makers to be modified to carry a torpedo, and was on *Hermes* by October. It was lost with the *Hermes* when that ship was torpedoed on 31 October 1914.[16]

RNAS 89 & 90 Short Admiralty Type 81 Folder Seaplanes

These two three-bay 160-hp Gnomê rotary engine folder seaplanes were ordered under Contract CP34603/13 but cancelled by January 1914.

S.82 to S.85 – Short Admiralty 81 Type Folder. RNAS Nos. 119 – 122

These machines were popularly known as Short Folders. Four Short Admiralty 81 Type Folders were purchased under Contract CP53783/13/X/16128 - Nos. 119 to 122. These were three-bays with a larger low-aspect-ratio fin with curved leading edge, a feature of many later Shorts. The wings of these 'production' aircraft could be folded from the cockpit. An identification feature was the lack of a fairing behind the rear cockpit that had featured on the earlier Folders. No. 119 was delivered to Grain and tested and accepted on 28 May 1914. No. 120 was tested and accepted at Grain on 15 June 1914. A copy of the report by the test pilot, Gordon Bell, is located in the machine's Log Book and reproduced hereunder.

Nos. 119 to 121 were sent to Calshot to take part in the Naval Review in July 1914. On the 9th Nos. 119 and 120 *ran short of petrol considerably before they should have done. This points to Shorts not providing for sufficient consumption of his 160's.*[17] Due to problems with over-heating of the double-row engines 160-hp Gnomê rotary engines, Samson had a vertical exhaust stack added to the top of the engine cowl.

Sqn Cdr Arthur M Longmore undertook the testing of the concept of launching a torpedo and when Winston Churchill visited Calshot seaplane station on 15 June 1914, he asked the CO, Longmore, whether the torpedo-dropping experiments could be hurried up given the situation that was unravelling on the Continent. The Sopwith Special Tractor Seaplane No. 170 was at Calshot. It had been designed on the ideas of Lt Douglas H Hyde-Thomson, a torpedo officer, together with Cdre Murray Sueter to carry and launch a torpedo. Longmore knew that No. 170 was having great difficulty in flying with any sort of load and told Churchill *that if I could retain a certain Short 160 Gnôme seaplane, I would undertake within a short time to carry and successfully release a 14-inch torpedo which weighed some 900 pounds. The retention of the Short was agreed and on July 28th I succeeded in getting this machine off the water with the torpedo slung below and in launching it for its run. It all worked out well but, of course, it was a stunt.*

No technical expert could pass the factor of safety of the machine so loaded, petrol was only sufficient for about half an hour and there was no passenger carried. … However, this preliminary success was very encouraging to the torpedo expert, Flt.-Lieut Hyde Thomson, who had done such excellent work in adapting the torpedo for release from the air and in designing the torpedo carrier which fitted on the seaplane under the fuselage between the two floats.[18]

Answering a call from the Admiralty for an aeroplane to carry a torpedo, Oswald Short was asked if the standard Short seaplane could carry a torpedo and while he thought that it could, he would have to check first with his brother, Horace. Working rapidly Horace designed modified cross bars on the floats and they were made and fitted in one night.

On 21 July 1914, a dummy torpedo was lashed on to the chassis of No. 120 and the machine tested for lift, *the dummy was 250 lbs short of full load but the machine got off the water correctly and it is thought that with alterations she will do it Telegraphed ICA accordingly.* On the 23rd, No. 121 arrived at Calshot from Grain and she was placed in No. 1 Shed. *Shorts man arrived at 4.30 pm to carry out necessary alterations to No 121 to fit her for torpedo.* This work took place over the next couple of days, the Shorts mechanics altering the chassis. The engine was stated to be *getting on well.* The afternoon of the 27th saw Lt Ross take No. 121 out for a practice flight. *Afterwards tried lifting torpedo with no success as engine not full out. Worked late on engine. Torpedo remains firm in spite of buffeting.*[19]

On the 28th Longmore launched a torpedo from Short No. 121. He recorded in his Logbook that he *Commenced*

Record of Acceptance Trials of Seaplane No.120

Maker: Messrs Short — Engine Type: Gnome 160
Type: Tractor folding wings.
Pilot: Mr G. Bell
Passenger: Mr Fairey. — Weight 385 lbs.
Overweight 85 lbs, Anchor 40 lbs.
Fuel: Petrol 49 galls. (11 galls deducted for overweight)
Oil: 14 Galls.
Weather: b.c. N.Fly 3-4

Hour endurance trial
Commencing 1.40 pm — Finishing 2.40 pm.

Climbing trial
Commencing 1.50 at 50 feet — Finishing 2.3.30 at 2050 feet.
Intermittent heights — feet after — mins — secs
(Note: These are blank)

Remarks. 2000 in 13½ mins.
Specification 3000 in 30 mins. Climb considered quite satisfactory & accepted.

Speed trial (Measured mile off Maplin sands)

	Time	Speed
	Time	Speed
Down wind	Down wind	
Up wind	Up wind.	

(Note: These were also blank)

Estimated speed. Estimated at 59 knots. Minimum at 39 knots. Both approximate

Remarks.
General remarks:
(1) During the climbing test the seaplane was held at an angle of 8° beyond normal horizontal angle as a test for a long period & looked over steep from the ground.
(2) Against the wind she appeared rather slow & her speed not exceed 59 knots. On test with No 82 this speed is confirmed.
(3) She has a big range of speed and lands slowly.
(4) She is fitted with a very strong & good sprung chassis which has proved excellent in practice in rough weather (5 ft waves).
(5) The side floats are considered to have too little buoyancy to be of much service.
These seaplanes handle well on the water under their own power but a larger and stronger waster rudder is desirable for towing purposes etc.

Note: This form was typed with blanks for the data to be filled in at the time of the trial – these remarks are shown in italics. The results of this test were entered into the machine's Log Book under 'Details of Machine.'

Source: Copy in Short No. 120 Log Book located at TNA AIR1/2056/204/412/120.

work getting machine out & standing by for orders. I took 121 out with torpedo managed to get off the water & to fire it successfully, torpedo made a good run, alighting afterwards was difficult as water was hard to see.[20] On 13 August, Flt Lt Ross lifted a torpedo on No. 121 during a one hour flight. The seaplane was not able to climb fast and the seaplane had eventually to be landed downwind in a force 5 wind, but the landing was successful *and the whole experiment considered as highly satisfactory. The piloting was of the highest class.*[21]

In 1926 Oswald Short wrote to the Air Ministry in relation to the claim by Sueter and Lt Col Malone in respect of Torpedo Carrying Aircraft. Oswald had to rely on memory as Shorts was unable to find any letters or instructions relating to the matter. His recollections were that in July 1914, *a 'Short' seaplane (Fitted with a 160 H.P. Rotary "Gnome" Engine) then in the Royal Naval Air Service at Calshot Air Station, was altered by us in certain particulars and adapted to carry a small torpedo of some 750 lbs. weight. It successfully flew with this torpedo about four days before the outbreak of war.*

The alterations consisted of fitting new cross bars between the floats, the original straight cross bars being replaced with cross bars which were bent upwards into a semicircle to allow of the torpedo being attached beneath the cross bars where it would have a clear fall and yet have reasonable space between it and the water. The bending of the cross bars in this manner weakened them as struts and to strengthen them it was necessary to fit vertical struts from the centre of the bends to the fuselage above them. Several minor fittings were possibly altered in the course of this work of conversion. The torpedo was carried beneath the cross bars to which it was attached by steel straps, and these steel straps were fitted with a quick release catch which could be operated by the pilot.

After this successful attempt and later on during the early part of the War in 1915 and 1916 our standard type of machine, known as the "Short" 184, originally designed as a reconnaissance and bomb dropping seaplane, was fitted with a similar arrangement of bent cross bars so that a torpedo could be carried, and a considerable number of these machines were turned out so fitted.

So far as I remember the actual steel straps to support the torpedo were supplied to us by the Admiralty, and the quick release catches were supplied by Messrs. William Walker & Son, of Bradford.

Oswald had no knowledge of the part played by the late Lt Col Hyde Thompson in originating the idea of a torpedo carrying aeroplane. Shorts discussions were all with the then Capt Sueter, *Chief of the Air Department of the Admiralty. We were requested by this gentleman to call at the Admiralty and he*

Left: Folder Type C No. 816. Note the translucency of the clear doped fabric where the sun shines through the cockpit.

Below: No. 818 appears to have no white field for the serial on the rudder. Note the kingposts on the upper wing bracing the wing overhang. Delivered to Grain on 1 December 1914; it was to serve with HMS *Rivera, Engadine, Aurora,* and *Empress* before being deleted on 11 August 1915.

made known to us his desire that we should endeavour to take a torpedo into the air on one of our standard seaplanes. A discussion followed as to the ways and means of doing this, and it was finally decided that we should make the alterations to the cross bars and carry the torpedo underneath them as already set out.

In consequence of this we have assumed that Admiral Sueter was responsible for the broad idea of carrying torpedoes on aircraft, and here is no doubt that he helped with suggestions as to how this could be carried into effect during the discussions referred to. It is only natural that we as builders of the machine which was to be converted should be called in to decide what could and what could not be done with regard to such conversion.

Generally speaking the alterations to the first machine were of a simple nature and were carried out in a few days, the matter being urgent.

Oswald stated that the only credit that Shorts would claim for themselves was *that at such an early date of aircraft development and with the low engine power then available, the efficiency of our machines was such as to enable these heavy weights to be carried with reasonable performance and safety.*[22]

Experiments continued through 1914. Grain reported on 9 August 1914, that No. 120 had flown with two 100-lb bombs that were dropped at 500 feet. No passenger was carried and the machine was sluggish and difficult to climb.[23]

Right: Launching No. 818.

Left: A Short Type C folder on the beach. The marine pilot's clothing of the era - long leather coat, helmet and goggles, and life preserver.

Right: The same machine displays the bomb rack under the fuselage. In the background is Wight Admiralty 840 Type seaplane No. 840. Location possibly Calshot.

Flt Lt R.P. Ross made a flight in No. 121 on 13 August, and another on 21 September when he managed to become airborne and successfully dropped the torpedo in the Humber. A few other Short Folders were modified to carry the 14-inch 810-lb torpedo. A Minute dated 3 August 1914, noted that Shorts Nos. 82, 119, 120 and 122, all 160-hp Gnôme powered machines, were to be converted for the carrying of torpedoes. No. 82 appears to have been modified at Shorts in September 1914. No. 119 was sent to Shorts on 16 August 1914, to fit modified chassis for torpedo carrying. Nos. 120 and 122 had bomb dropping gear fitted in August.[24] However, the Admiralty H.M. Aircraft Lists for December 1914, shows none of these aircraft with a "T" for torpedo or a "B" for bombs.[25] Nos. 119, 120 and 122 were used as bombers, and No. 121, Longmore's aircraft, was modified for bomb dropping in February 1915.[26] Three Short Folders of the seaplane-carrier *Engadine*[27] were modified to carry torpedoes but no torpedo attacks were ever carried out by these machines. This leads to the conclusion that the 160-hp Gnôme Shorts were inadequate as torpedo aircraft.

Short S.88 & S.87 - Admiralty 135 Type Folder Tractor Biplane Seaplanes. RNAS Nos. 135 & 136.

A late 1913 design, the Type 135 was a two-bay floatplane with the upper wing extensions braced by struts from the lower wing. The ailerons were typical Short single-acting type. Problems had been experienced with the twin-row 160-hp Gnômes that tended to overheat and the two seaplanes had the Salmson radial water-cooled engines installed. No. 135 had a 135-hp Salmson single-row radial engine that was cooled by a large radiator mounted in front of the forward cockpit. No. 136 had a 200-hp double row Salmson engine. Salmson engines were known as Canton-Unnés after their designers, Georges Canton and Pierre Unné.[28]

Short Brothers delivered No. 135 to Grain on 28 August 1914, under tow by a steam boat from Queensborough Pier at 4:00 p.m. The engine was tested and the machine prepared for its acceptance flights. The seaplane went to the *Rivera* on 9 September to test the hoisting arrangement of the ship and for a test flight. On 12 September No. 136 was delivered to Grain prior to acceptance tests.[29]

Both took part in the Christmas Day raid on Cuxhaven. The smaller, lighter and less powerful No. 135 was lost in the raid.

No. 136 was damaged on 25 October 1914, when a 100-lb bomb exploded when it was alighting. Repaired, it was on HMS *Riviera* for the Christmas Day raid.

Specifications No. 135:
Span 52 ft; Length 37 ft; Area 530 ft^2; Weight Empty 3,050 lbs; Loaded 3,400 lb. Speed max 65 mph. Endurance 4.5 hrs.

Specifications No. 136:
Span 54 ft 6 in; Length 40 ft; Area 570 ft^2; Weight empty 3,000 lb; Loaded 3,700 lbs. Speed max 72 mph; Endurance 4 hrs.

Short S.108 to S.115. Type C Folder (Improved Admiralty 74 Type) Tractor seaplane. RNAS Nos. 811 – 818

Nos. 811 - 818 (S.108 – S.115) was a small batch ordered from Shorts at Eastchurch. These eight seaplanes had varied and interesting service lives. They were powered by 100-hp Gnôme engines and resembled the earlier type, but had their wing extensions braced by cables from king posts above the upper wing. No. 811 was delivered to Grain in October 1914, being test flown on the 29th, and subjected to speed tests over a measured mile off Maplin Sands on the 31st. No. 813 arrived in tow from Shorts, Rochester, on 5 November. Nos. 811, 812 and 135 were transferred to *Rivera* on 21 November 1914.[30]

Reginal John Bone recalled that *The authorities thought that the public needed encouragement and organised large scale air raids behind German lines ana on Ostend and Zeebrugge. For one of these my carrier* (HMS Empress) *was sent to Dunkirk to swell the total of aircraft-up so that a communique could be issued 'a thousand of our aircraft raided etc., etc.'*[31] The attacks by Cdr Charles R Samson's men at Dunkirk on Zeebrugge and Ostend led the Admiralty to plan a large attack on these two submarine harbours. Intelligence reports confirmed damage to at least three enemy submarines and it was decided *to keep up a constant bombardment of the Canals, with the view of destroying any submarines, tenders, motor boats, or other small craft that may fuel the submarines; power stations, etc.*

The first combined aeroplane and seaplane attack will tazz place about the 11th February, or first suitable day afterwards.

The following aircraft would be provided by:

Aeroplanes	
Dover	10
Eastchurch	6
Hendon	4
Dunkirk	4
Seaplanes	
Dunkirk	4
HMS *Empress*	4
Felixstowe	4
Dover	2
Total	38[32]

Aircraft and seaplanes from the UK were sent to Dover and Dunkirk. The morning of the 11th saw three aircraft attack; two landplanes and one seaplane only started due to the gathering storm. The seaplane was Short No. 812 flown by Lt Bone, who dropped one 100-lb bomb alongside the railway station at Blankenberghe. *Our poor little folder seaplanes carried one 100LB bomb and four 20lb bombs. With this load they clambered very slowly to about three or four thousand feet. We were given a target at the landward end of the mole at Zeebrugge and strict instructions not to attack unless we had*

The Following Aircraft Took Part in the Raid of 16 February 1915			
Serial	**Type**	**Crew**	**Details**
16	Avro 501	Johnston	Failed to find target.
50	B.E.2a	Lt Richard Peirse	Ostend. Attacked wharves – 2 x 20-lb bombs.
179	Avro 504	Lt Buter	4 x 20-lb bombs on heavy gun position west end Ostende.
643[2]		Gaskell	Failed to find target.
812	**Short Folder**	Flt R Bone	Railway line at Blankenberghe. 1 x 100-lb bomb.
813		The Hon Desmond O'Brien	Lost at sea.[1]
816		Blackburn	Engine trouble, failed to find target. Temporarily missing.
817		Flt Sub-Lt Thomas Spencer	Lost at sea.[1]
818		Moon	Could not get off due fog.
819	**Short 830**	Hewlett	Could not get off due fog.
873	Avro 504	Flt Lt Harold Rosher	Zeebrugge mole. 3 x 0-lb bombs.
875		Flt Edward Gordon Riggall.	Zeebrugge power station. 4 x 20 lb bombs. Failed to return, Riggall killed.
876		Lt John P Wilson	Slykens bridge on Bruges canal. 4 x 20-lb bombs.
877		Maude	Crashed into sea due engine failure. Picked up by HMS Excellent. Engine & cockpit section salvaged.
908	Bleriot XI	Fowler & Nickerson	Ostend harbour storage sheds. 3 x 20-lb bombs.
1001	Avro 504B	Lt Edward Osmond	Dropped 4 x 20 lb bombs on mine-sweeper off Ostend.
1002		Lt Meates	Dropped 3 x 20-lb on battery on Eastern side Ostende Harbour.
1003		W Cdr Arthur M Longmore	Crashed on take-off.
1004		Flt Lt Sidney V Sippe	Had to return due to engine trouble.
1216	Bristol T.B.8	Haskins	Engine problems. Dropped bomb load in sea.
1223		Bettington	5 x 20-lb bombs on guns at Middelkerke.
1224		Dawson	Slype canal. 2 x 20-lb bombs.
1225		Robertson	Engine would not start.
1241	Maurice Farman	Marix & Lt Collet	3 x 100-lb bombs on heavy gun battery, Mariakerke.

Notes: 1. Sturtivant & Page, while recording the loss of the two machines with the loss of the two pilots, do not confirm the serial Nos. as the machine flown by O'Brien and Spencer, however Hobson does.

2. Serial No. 643 belonged to an RFC B.E.8.

Source: Oliver, J. *Samson and the Dunkirk Circus*, Oliver, UK. 2017. PP176-177.

reached five thousand feet. On this account of the accuracy of the German AA.

On the day of the raid our four seaplanes took off but snowstorms caused us to lose touch with one another. In consequence my approach to the target was at 1000ft and I was surrounded by AA bursts – my instrument board was smashed and a fragment lodged in my handkerchief which was -in the fashion of the period – pushed up my left sleeve.[33]

The weather for the remainder of the day was not suitable for flying. That night, as soon as it was feasible, Samson flew Maurice Farman No. 1241 to attack guns at Middelkerke and Blankenberghe with 100-lb bombs. He achieved this end despite the weather and fog banks.

The 12th saw a raid by 18 aircraft, which included Short Folder No. 816 and 830 Type No. 819.

Another raid was arranged for the 16th. The French Military Aviators also took part *sending 8 aeroplanes* (Voisin) *to drop bombs on Ghistelles Aerodrome.*[34] This time six Short seaplanes took part. The attack took place at 1500. Four pilots were reported missing afterwards, three being killed.

The Netherlands Ijmuiden's steam pilot boat No. 6 found the wreck of a seaplane off Ijmuiden on 4 March 1915, and took it in tow, but lost it again. It was later brought in by torpedo boat. The wood and aluminium work was destroyed. The aircraft had a Gnôme engine and an English propeller No. 287R and three gauges on board.

Left & Below: Short Admiralty 135 Type Folder No. 136 on 27 April 1915, Kephalo Bay, Dardanelles. Sent out on HMS *Ark Royal* in January 1915, No. 136 made its first reconnaissance flight over the Dardanelles on 19 February. The incident shown here was the result of a spotting flight for HMS *Triumph* on a Turkish battleship in the Narrows. The aircraft was hit by Turkish ground fire and a port chassis strut was shot through, collapsing on alighting. Crew was Lt E.H. Dunning with Lt W Parke. Back in the air by 28 May 1915. Wight and Sopwith seaplanes in the background.

Left: Capt Kilner aboard No. 136 on the hoist. Note the flaps hanging down from the lower wing.

Right: A Short 135 Type with wings folded and connected to the ship's hoist. Note the two men on the front of the floats. It is apparent that this was done when the machines were being hoisted, apparently in order to get the centre of gravity correct for hoisting. The padded poles used by men on the ship to prevent the machine striking the vessel during hoisting may be seen. The small armed ship in the background is an armed steam picket boat from one of the big ships.

Left: The observer on the starboard float of this taxiing Short 135 Type is, presumably, waiting in this precarious position for the pilot to cut the engine so that he can pick up a tow rope.

Right & Page 94: Short Folders Nos. 119, 121 and 122 were transported to Durban, East Africa, by the liner Laconia, arriving on 14 April 1915. Nos. 199 and 122 are shown after being erected on the Durban dock. The 'chimney' exhausts have not been installed yet; the red ring type of RNAS markings is carried on the wings. The large packing crates for the seaplanes are on the dock. All the machines were found to have been damaged in transit. (cont.)

This was most likely Short No. 813 or 817. Both these were lost on 16 February 1915, on operations from Dunkerque. The pilots were Flt Lt the Hon Desmond O'Brien and Flt S/Lt Thomas Spencer respectively. It is most likely that the wreckage was from one of these seaplanes. The bodies of the two men were never found.[35]

The Christmas Raid on Cuxhaven

On 7 August 1914, the fast cross-Channel steamers *Engadine, Rivera* and *Empress*, were sequestered by the Admiralty for conversion to seaplane carriers. The first two went to Chatham Docks where they had structural alterations undertaken to enable them to carry out their new role.

Only essential work was carried out to enable planes to be put on board and hoisted out. Canvas was used to shelter the planes, which it did very inadequately. The Empress was used to carry seaplane parts and stores at first, but in October she too was sent in for the same alterations and additions.

It is quite clear that there was at this time no thought of using these ships with the fleet. It is possible that the Admiralty had decided that, with the unfortunate results of attempted fleet co-operation in the past, there was no useful purpose to be served by adding any of these ships to the fleet. The great threat to be faced at this time were the Zeppelins and submarines.[36]

It was proposed to use these carriers primarily to go to the enemy's coast and attack the Zeppelins in their hangars. From the first, the Admiralty took an aggressive stance with their air service. The Cuxhaven Raid of Christmas 1914 was an example of this. On 24 December the seaplane carriers *Empress, Engadine* and *Riviera* were escorted by the Harwich Force under Adm Tyrwhitt, comprising light cruisers *Arehusa* and *Undaunted*, together with eight destroyers of the Third Flotilla. Ten British submarines were to operate close into the German coast while the Grand Fleet waited further out in the North Sea where it was hoped that the German fleet could be lured into a major engagement.

The raiders sailed from Harwich and at 6 am on Christmas Morning they were in position some 12 miles north of Heligoland. A co-operating force of two destroyers and ten submarines had been sent to repulse and enemy attacks and to pick up the crew of the seaplanes if needed. The force had been sighted at dawn and the whole North Sea coast had been alarmed by the watchers on Heligoland Island.

The three seaplane carriers had a total of nine Short seaplanes on board and they were all hoisted over the side. No. 122 under Flt Cmdr Gaskell and No. 812 under Flt Lt Bone were unable to get off and were hauled back in. The other seven all got off and were all Short seaplanes:

Short Admiralty 81 Type Folders
No. 119. Flt Cdr Robert Peel Ross. *Engadine.*
No. 120. Flt Lt Arnold John Miley. *Engadine.*

Short Admiralty 135 Type Folder
No. 135. Flt Cdr Francis Esme Theodore Hewlett. *Rivera.*
No. 136. Flt Cdr. Cecil Francis Kilner with Lt R Erskine Childers, RNVR, as his observer. *Rivera.*

Short Type C Folder
No. 811. Flt Lt Charles Humphrey Kingsman Edmonds. *Rivera.*
No. 814. Flt SLt Vivian Gaskell-Blackburn with CPO James W Bell. *Empress.*
No. 815. Flt Cdr Douglas Austin Oliver with CPO Gilbert H.W. Budds. *Empress.*

Three seaplanes returned to the ships. Three pilots landed near the Submarine E.11 near Norderney Gat and were taken aboard, their seaplanes then being sunk.

Hewlett was missing. *His machine was seen in a wrecked condition about eight miles from Heligoland, and the fate of this daring and skilful pilot is at present unknown* reported *The Aeroplane* about the raid.[37] Hewlett wrote that *the engine began to misfire slightly and the revolutions dropped and I noticed that the oil pulsations in the sighting glasses were becoming very feeble. In a few minutes the glasses emptied and the engine was obviously overheating. I therefore looked round to see if any ships were in sight and saw only a small trawler, towards which I glided and circled to make certain of her nationality. Seeing she flew the Dutch flag I landed by her.* The trawler could not provide him with any oil so he waited from 10.00 a.m. till noon. He attempted to take off and search for British ships but was unsuccessful and returned to the trawler. He *waited until dark in hopes of sighting some ship and as the wind was getting up, I decided to sink the machine and remain aboard the trawler Maria van Hattem. Holes were punched in both main floats, tail compartment and by revolver shots in the tail float and the machine settled by the bow and finally sunk with all but the tailplane and rudder.*[38] Hewlett reported that No. 135 had problems with its oil supply during trials and that he had secured the correct adjustment on the machine, however when he alighted by the trawler on Dec. 25th *I noticed almost at once that the set screws were slack and that the adjusting screws in about half an inch, therefore too much oil had passed to the engine and not enough returned to the tank.* As the heads of the screws were in their correct relative position to each other he did not notice that they were incorrectly set before his departure. *I beg to submit that I deeply regret that I should have caused so much trouble and should have necessitated the loss of a seaplane of the Naval Air Service.*[39] He was interned but eventually released as a shipwrecked mariner.

The fate of the seaplanes was as follows:

Short Admiralty 81 Type Folders
No. 119. Came down and towed in by HMS *Lurcher.*
No.120. Came down near Norderney Gat lighthouse. Taken in tow by submarine *E.11.* Bombed by German airship so it was abandoned and sunk by gunfire.[40]

Short Admiralty 135 Type Folder
No. 135. Forced to alight. Fate unknown.
No. 136. Recovered by *Riviera.*

Short Type C Folder
No. 811. Recovered by *Riviera.*
No. 814. Abandoned near Norderney Gat and crew rescued by submarine *E.11.*
No. 815. Abandoned near Norderney Gat and crew rescued by submarine *E.11*

F.J. Rutland considered that, *strange to relate, this was probably the most successful attempt made with seaplanes during the whole war.*[41] A mist over the target at Wilhelmshaven made finding the sheds difficult. The attacking force was twice attacked by Zeppelins but no hits were registered.

Zeppelins *L 6* and *L 5* from Nordholz took off to attack the raiders. *L 5* sighted three seaplanes over the River Weser steering for Nordholz and sent a warning to the base. The weather now turned against the attackers; a low, thick fog drifted in from the sea. At 9.20 a seaplane was heard over Nordholz. The plane came into view and was fired on. It came down to about 1,000 feet and made for the revolving airship hangar, then turned and made for the gasometer holding a million cubic feet of hydrogen. It dropped two bombs that missed the target. As no pilot claimed dropping his bombs on a target, the identity of the aircraft involved is unknown.

L 6 sighted two light cruisers and eight destroyers and three *steamers with funnels painted black,* failing to recognising them as aircraft carriers. *Empress* was having condenser trouble and lagging behind the force steaming away from the coast. Gaining height *L 6* made to attack the straggler. *L 6* dropped one 50-kg bomb at *Empress* that missed by at least 100 feet. The other three bombs were dropped as ballast to get the airship to rise after it came under fire from the ships. British accounts have *Empress* sailing under a rain of bombs, the attack only ceasing when the Zeppelin was hit by a 6-in shell from the cruiser *Undaunted.* Inspection of the Zeppelin the next day revealed nine holes from rifle bullets in the gas cells. Despite losing gas, the ship arrived back at Nordholz light and gas had to be vented in order to land the ship.

L 5 came upon the submarine *E.11* just after the last two seaplanes arrived to be taken aboard. The submarine just managed to take the two airmen onboard and submerge when two bombs were dropped nearby. *L 5* then went west and found the raiding force and followed them before returning to the three seaplanes that were shot up by his crew. The Zeppelin descended to a height where he could read their serial numbers, Nos. 120, 814 and 815, as well as record their markings – a red circle on the top of the wings and a Union flag underneath.[42]

As a result of the raid the Germans increased the protection of their airship bases. 88-mm and 37-mm anti-aircraft guns were installed and the hangars with their large reflective doors, that could be seen from 12 miles away, were camouflaged.[43]

Due to the success of this raid, it was decided to fit these ships as permanent carriers. The large structural alterations involved led to the docking of the ships for a period of approximately three months. In addition, two ships were taken over from the I.O.M. Steam Packet Co, the famous *Ben-my-Chree* and the *Vindex.* All these ships initially carried 150-hp Canton Unné Shorts, the *Engadine* carrying five Sopwith Baby seaplanes. All had a steel hangar with electric cranes upon either side to hoist planes in and out. A small workshop with lathes, etc., was fitted below. There were four Slingsby trolleys on which the planes were placed. They were of special design and had a screw braking system that bore down on the deck. Hence securing chains were not necessary unless a certain degree of roll was reached. This enabled *the planes to be moved in and out without danger of a sudden lurch of the ship causing the plane to take charge.*[44] The Harwich Force had noted that the Zeppelins would be about whenever the weather permitted and so the Sopwith fighters were carried in order to attack the Zeppelins. The ability of the airships to climb rapidly to heights that the Sopwith could not reach rendered the single-seat fighter a deterrent, rather than an offensive weapon, as the Zeppelin would have to operate at a higher altitude. (See Chapter 13 for the use of the Sopwith floatplanes in Operation G.)

No. 136 at the Dardanelles

Col F.H. Sykes, RFC, attached to General HQ, Mediterranean Expeditionary Force, presented a Report to the Admiralty on the 'Aerial requirement of the Naval and Military Forces at the Dardanelles.' He noted: -

Conditions generally under which the units of the R.N.A.S. have carried out their work during the Naval and Military operations at the Dardanelles have been exceptionally difficult. Many of the units had only been recently formed and had no experience of service away from good sheds and other conveniences. The heat, flies and dust have rendered the work of Officers and men on shore very laborious. Though the atmosphere is wonderfully clear for aerial observation, the glare from the sea and the large areas of sandy country causes severe strain to the eyesight of observers – more especially as but few of these had received previous training in their work...

The Seaplane unit of H.M.S. "ARK ROYAL" *has sometimes secured good results, ... the machines in use cannot be said to have proved suitable to assist in the operations involved.*[45]

It had been announced in August 1914, that the mother ship for seaplanes that was under construction by the Blyth Shipbuilding Co, was to be named the *Ark Royal.*[46] The *Ark Royal* proved itself a suitable design for the work for which she was intended, however an empty collier, the SS *Penmorvah,* was obtained for use for the storage of spares and for erection and repair work. Without the help of such an auxiliary or shore base, the *Ark Royal* would not be able to

They were repaired and back on the Laconia by 18 April, together with their support personnel.

Above: Flt Lt John Tulloch Cull and LM Boggis about to take off in No. 122 off German East Africa. No. 122 was used for reconnaissance of the Rufiji Delta in May 1915. The machine could not climb well and could not climb beyond 1,200 feet. The age of the seaplanes and the climate conditions rendered them inefficient as described in the main text.

Right: The SMS *Königsberg* before the war.

Above & Right: The *Königsberg* lies wrecked but the crew were not defeated. The guns were removed and made into field pieces and they joined the German Army in their long fighting retreat from the British.

work with more than six or seven seaplanes.

On 9 May hostile submarines made their first appearance. The German U 21 sank the British battleship *Triumph* on the 25th and, two days later, the *Majestic.* Layman considered that this *little submarine had a strategic impact out of all proportion to her size* in the absence of any effective anti-submarine weaponry in the theatre.[47] *Ark Royal* was sent to Kephalo Bay, where she remained until the end of October. The prevailing NE wind caused a considerable sea in the bay and would have prevented flying from the ship; accordingly, a shore base was started at Aliki Bay which has a Southern exposure, and from it, regular flights were made for spotting, reconnaissance and photography on the Gallipoli Peninsula and submarine patrol in the vicinity.

Sykes noted that the Wight and Sopwith machines were not satisfactory. The 200-hp Canton-Unné Short *are given the best character and were not that their engines have given a great deal of trouble would probably afford better results.* The engines had to be flown at full throttle which led to rapid deterioration.

Overall, Sykes thought that on the whole *the results gained by seaplanes are disappointing and experience points to the fact that it is wasteful to employ seaplanes for work that can be carried out by aeroplanes. The engines employed are not so reliable, the machines are more frequently prevented from flying by weather conditions, and they are more difficult to keep in efficient order. It is exceptional for a seaplane carrying a passenger to be able to get higher than 3000 feet. This precludes flying over the enemy – particularly as in the area of Gallipoli peninsula considerable air disturbances and down currents are stated to be prevalent.*

The seaplane therefore is very heavily handicapped in regard to such work as observing gunfire unless the hostile trenches, guns or other targets are close to the coast-line. They are also practically of no use for photography. The observation of ship's fire on to conspicuous targets such as villages or large forts can of course be carried out by seaplanes.

A seaplane-carrier, such as H.M.S. "ARK ROYAL", would appear to be best used when, owing to the distance from friendly land a landing ground for aeroplanes cannot be obtained and a carrier can proceed to the nearest point and fly her machines from there. This has been done by the "ARK ROYAL" on various occasions. Nor must it be forgotten that in event of the Fleet getting into the Sea of Marmara, it will be probably some time before a land base can be established, and seaplanes would probably be valuable for searching out enemy ships etc.

It will be extremely interesting to see whether the seaplanes of the 'BEN-MY-CHREE" are to effect a great concerti or moral result by means of their torpedoes. One has not yet been fired against the enemy.[48]

Seaplanes did good work here on the 18th and 19th and *they will be invaluable in the later stages* of the campaign. More machines were requested to replace wastage. The CO of the *Ark Royal* asked for Short No. 161, another Wight and three 135 or 200 horsepower *Canton Shorts* with a liberal supply of spares. *Ark Royal* advised that four seaplanes could not be used under the *conditions prevailing here.*[49] Sueter minuted the 1st Sea Lord that arrangements were already made for the despatch of the Wight No. 176 *(200 Canton-Unne)* and Short No. 161. *This will bring "ARK-Royal" up to a total of 8 seaplanes.* It was not possible to send more Canton-Unné seaplanes at the time without upsetting the arrangements made for other seaplane carriers.[50]

On 27 April 1915, the Turkish battleship *Turgud Reis*[51] attacked the *Ark Royal.* Flt Cdr E.H. Dunning with Sub-Lt W Park took off in the Short to direct fire on to the Turkish ship, but it steamed out of range. Coming back over the lines they came down to 2,500 feet to search for the enemy's batteries, they were fired on and one of the float struts was shot through. On alighting, the strut broke. Once repaired No. 136 was again ready for action but the forward struts buckled and the nose was immersed with detrimental effect to the Salmson engine. The official historian, H.A. Jones, recorded that although she was tried a few times more, No. 136 faded from the picture.[52]

The Destruction of SMS *Königsberg*

In 1886 Germany annexed 384,000 square miles of East Africa to give the newly united nation a new colony as it worked to establish its place in the sun. The harbour city of Dar-es-Salaam, the home of the old sultans of Zanzibar, became the trading capital of the new colony. In June 1914, the arrival of the German cruiser SMS *Königsberg* under *Fregattenkaptän* Max Looff was the cause of great celebrated. The cruiser was only seven years old when war broke out in Europe and, stationed at Dar-es-Salaam, was more than a match for the old British cruisers that the Admiralty had in the Cape of Good Hope Squadron. The pre-war situation had not required that the best and latest ships be sent to a station that covered an area that extended as far west and north as St Helena, as far east and north to include the whole of East Africa and its outlying islands.

Königsberg had ten 105-mm guns, two 45-cm torpedo tubes and displaced 3,350 tons. The European situation was known to the opposing ships and the evening of 31 July 1914, Adm Herbert King-Hall had dispatched HMS *Astræa* and *Pegasus* to watch for the *Königsberg* at Dar-es-Salaam, and was steaming alone when his flagship, HMS *Hyacinth*, came upon the German warship. The two ships passed within 3,000 yards of each other. *Hyacinth* tried to follow but the British ship was slower and had poor coal and was short of even that, so the German vanished into the night. Four days later the encounter would have been different. As it was the German cruiser went on to sink Allied shipping and tie up a vast resource as the threat of the *Königsberg* as a raider to Allied commerce in the Indian Ocean was always greater than its actual effect, but could not be ignored. This was more so after the *Königsberg* caught *Pegasus* in Zanzibar

Harbour on 20 September 1914. The British ship was at anchor cleaning its boilers and did not stand a chance. She was sunk with the loss of 31 lives and 55 wounded. Cdr John ALexander Inglew struck the colours to avoid further bloodshed.

The coast of East Africa was full of small harbour, creeks, islands and little-known areas where supply ships could rendezvous with the raider. All this had to be patrolled in order to find the raider's location. The *Königsberg* had taken shelter up the Simba Uranga channel of the Rufiji Delta, south of Dar-es-salaam. The delta contained five major branches of the river that split the area into hundreds of square miles of marshland, mangrove swamps and small islands. Ships that should have been protecting convoys were called back to hunt for the raider. Eventually the British decided that the *Königsberg* might be up the Rufiji River. This was going to be difficult to determine as, according to existing maps, the waterways were too shallow for a ship of the *Königsberg* size, except, perhaps at the spring tides. For the British it was impossible to go up the river and equally impossible to see anything from the outside. When discovered by *Chatham* on 30 October, the raider was anchored with the *Somali* that had a cargo of grain and wood for the *Königsberg*. *Chatham's* master, Captain Sidney Drury-Lowe, was frustrated by the situation and on 7 November took his ship within two miles of the shore and heeled it over for greater elevation and opened fire at a range of about 13,000 feet. A hit on the *Somali* caused it to burn for two days, depriving the *Königsberg* of much needed supplies including coal. The cruiser was further up the river and out of range.

In order to attempt to block her exit two colliers were sunk in one of the many channels comprising the delta by the *Chatham*, however, many still lay open. This attempt to block the channel with a sunken ship was partially successful. Now the plan was to locate and destroy the *Königsberg*. The RN approached Herbert Dennis Cutler who was operating a pair of Curtiss F-Boats giving flying exhibitions in Durban. The Curtiss boat was old, lacked spares, and Cutler did not have a mechanic. Royal Naval Reserve midshipman Arthur Noel Gallehawk, who was serving in the *Kinfauns Castle*, the ship that brought the flying boat up from Durban, was assigned to assist Cutler, who was given a temporary commission as a Sub Lt in the RN. He was to carry out an aerial reconnaissance of the Rufigi.

Cutler and Gallehawk managed to get the Curtiss into a fit state to carry out the proposed flights. On 22 November 1914, Cutler carried out his first reconnaissance at 7 am. The *Konigsberg* was located lying further up the river, as far up the river as the ship could manage to sail, about seven or eight miles in a straight line. Flights were necessary every day to keep track of the raider. Although blockaded in, the exploits of the *Emden* showed what a raider could do if let loose in the trade routes. On 10 December 1914, the flying boat suffered engine failure and had to alight on the river where Cutler was taken prisoner. Gallehawk led an armed party to recover the boat, Gallehawk himself succeeding in making a line fast to the boat that had been beached, and it was recovered under heavy German fire. It was eventually taken back to Durban where it was displayed in a museum.[53]

Altogether, this little episode, although not generally known, was a brilliant feat, as the two officers concerned had none of the usual and necessary facilities in the way of mechanics, spare, etc., but despite these and other difficulties managed to get the boat flying and to carry out the reconnaissance, which only failed to achieve its purpose through bad luck, while the salving of the flying boat was carried out in a skilful and gallant manner.[54]

In December, the C in C East Indies requested that a seaplane with bombs or a light armed vessel be sent to overcome the stalemate. The Admiralty decided to send out a small seaplane squadron, Expeditionary Squadron No.4, under Flt-Lt John Tulloch Cull with Flt-Lt Harold E Watkins and a party of 18 men[55] with two Sopwith Type 807 floatplanes (Nos. 920 and 921). Cull was ordered on 7 January 1915, to report to Calshot to arrange for two Sopwith seaplanes to be shipped for the Flight.

Anticipating trouble with the airscrews in tropical climates, two representatives of the Lang Propeller Co of Weybridge were taken out. CPO Jack Hugh Stewart Sprott, who was one of the directors of the company, and Leading Mechanic Burgess.[56] CPO Mobsby was a Sopwith employee and joined the expedition in a civilian capacity at 48 hours notice when it was finally decided to take out Sopwith machines. Short 'Canton Unne' Folders originally intended for the expedition were not available in time. Sailing on the P & O SS *Persia*, the passengers were told that the men and aircraft were to establish a training school at Bombay.

On arrival at Bombay on 8 February, the two aircraft were unloaded, taken to the dockyards, uncrated and erected and test flown. They arrived at Nyororo Island on 20 February 1915.[57] The first machine declined to leave the water with a crew of two, petrol and two 16-lb and two 50-lb bombs. It was proposed to bomb the *Königsberg* but the machine could barely reach 1,500 feet with only the pilot and fuel for one hour.

Trials were carried out here with the utmost secrecy but both engines and machines gave considerable trouble in the tropical climate. Sopwith No. 921 crashed to a wreck the first week with a total flight time of only 2 hours, leaving only one machine for the proposed work. In view of this the C in C Cape wired the Admiralty that nothing could be done until better machines were supplied. In reply, the Admiralty sent three Short floatplanes towards the end of March. These arrived at Durban on the latest Cunard liner, the *Laconia*, an armed liner from Cape Town, about the middle of April 1915.

As soon as the cases were opened, it was realised to our disappointment that this second lot of machines were in one way old friends to all of us, as they had been in use at the Isle of Grain previous to out departure. In addition they had not benefited by their passage out, and were generally in a bad condition, so that a great deal of work was necessary before they could be in any sense called fit for flying.[58]

The Short seaplanes turned out to be three Short Folders (Nos. 119, 121, and 122) rather than new machines. The engines were also in poor condition and had to be reconditioned before use. This occupied the detachment on the cruise back to Nyoroco Island, arriving on 23 April when flight trials were commenced. The Shorts were old, but certainly an improvement on the Sopwith floatplanes.

Watkins recorded the following of his first flight in the Short; - *I went up to test the machine, and having climbed with a passenger to the maximum, about 1,200 feet, I felt something wet blowing on me and realized it was petrol. I immediately switched off and started to glide down. When we landed I was soaked with petrol and the tank was empty. When the tank was removed we found 25 rust holes.*[59]

Actually, this was Watkins second flight in No. 119. On the 27 April, he went up with Cull for *Instruction with dual control – machine climbing excellently. 500' obtained though not trying to climb.* On his second flight this day with AM Boyce as passenger, he recorded: -*machine climbing very well – heavy rain in vicinity & air saturated. 800' in 14 mins. Bad leak in petrol tank forced landing. (Bottom of main tank eaten through with rust – Sopwith tank being substituted.)* [60]

On Sunday the 25th, Flt Lt Cull took AM Ebenezer H.A. Boggis[61] as observer and took off in No. 122.

The machine was not climbing very well and a height of only 1,200 feet could be reached. Nothing could be gained, however, by further climbing, so the machine was headed for the Delta, entering by the main southern mouth.

The Delta formed an area of some 100 square miles with five main channels but hundreds of branches and islands. The channels were in most cases impractical for navigation for any but the lightest draught boats as they are full of constantly shifting sand and mud banks *and there can be little doubt that the "Koenigsberg" which was serving in these waters prior to the war had taken the precaution of doing some thorough survey work against 'the Day,' so that she was able, when matters became dangerous, to take advantage of a tide and make her way up to the head of the Delta.*

On the seaplane crossing the entrance of the Delta, a down bump was experienced that dropped the machine to about 800 feet and it came under rifle fire that was to be *experienced all the way up the channel.*

The enemy having been reported to the south of the delta, were not spotted until the machine, having arrived at the head of the Delta, turned and flew north for some minutes. The "Koenigsberg" was then observed lying in practically the most western of the numerous channels... She looked as though she had been newly painted, her side screens and awnings were spread, smoke was issuing from her funnels, and in general she was looking very spick and span.

No sign was observed of any attempt to camouflage the ship by means of branches of trees as reported in the English press. The machine by now had arrived within range of the "Koenigsburg", and on her present course would have passed about three quarters of a mile on her beam, travelling from stem to stern. As the machine was not, however, flying very well, and everything of interest had already been noted, the machine was turned just before coming abreast of the "Koenigsberg", and headed out to sea again. As she turned, the "Koenigsburg" opened fire, and for a first attempt made some very pretty shooting, shells bursting just astern of the machine, which, however, was soon out of her range. The machine left by the same entrance as she entered, and was speeded on her way with a considerably reinforced volley of rifle fire, German officers and Askaris being easily distinguished from the height of 600 feet, firing for all they were worth and in no way disconcerted by pistol or Verys pistol fire from the machine.

From now onwards the engine started failing gradually, and packed up altogether when within 6 or 7 miles of the anchorage where the ships were lying. On landing, it was found that a rifle bullet of about ½" bore had entered between the observer and the engine, at a very flat trajectory, tearing away the air intakes and succeeding in practically closing the main oil pipe. This latter was naturally the cause of the engine failing, and several blued cylinders resulted. [62]

The floatplane had to be towed home, the photographs developed and a report made to the Admiral. *The CinC expressed himself very satisfied.*[63]

Further reconnaissances were carried out in which the Flag Commander Hon. R.O.B. Bridgeman usually went as observer, his interest and keenness in our flying operations being a great incentive to everyone to make the best out of what was available.

Nothing of particular interest occurred in these reconnaissances as the machines were not climbing well enough to be sent often into the Delta, and were usually flown within easy reach of the sea, this being quite far enough in to notice any change in the enemy's position. [64]

On 5 May, Watkins and AM Gregory in No. 119 were on a *Reconnaissance down coast. Machine flying too close to land & got heavily fired on from Ras & Wava – rudder carried away for some reason & forced landing across wind – capsized – broken up – Pilot & observer 3 hours in water until rescued by Lieut Cull in No 122.*[65]

The machine came down just off the enemy's shore. Cull and Leading Mechanic Boggis went looking for the missing aircraft and found the wreckage just in time as a moderate sea was rapidly breaking up what was left of the aircraft, the pilot and observer clinging to one of the floats. No craft was in the vicinity to assist so Cull alighted and took the pair onboard, then taxied until he found a patrolling whaler. This was sent off to get a second whaler so Cull could off-load his pair of passengers while the first whaler then returned to the wreck to try and salvage the engine.

After strenuous efforts in the failing light a line was made fast to the engine and it was buoyed for the night. The next day the *Chatham* attempted to reach the wreck as the whaler could not raise it. Taking a 5,200-ton ship into these waters was a challenge but aeroplane engines were scarce in this theatre and the attempt had to be made. The big vessel made its way slowly to the wreck and the ship tried to raise the engine that had sunk into the mud. Unfortunately, *Chatham* fouled the line with her propeller. After fruitless attempts by the divers to find the engine, the *mud was very thick and machine upside down with the nose in mud* the task was abandoned as the tide was turning and the *Chatham* would soon be aground.[66] No. 122's floats were damaged during the night when it was in tow by the motor boat *Fly*. Three ribs in the elevator and rudder were broken when getting into the Laconia. *As no spares have been sent Elevators & Rudder* (have to be) *rebuilt.*[67] So, after all the trouble taken to recover the engine, it was to no avail.

The following report had been sent to the Admiralty at the beginning of May:

Machines not climbing well. Average speed attained 60 m.p.h. Engines giving 1140 revs. Climbing best when humidity great. Consider great deal of slip present. Glue not holding on any propellers. Wood for only one more propeller. Has indiarubber tubing been sent? As all ours perished. No fabric and few spares left.[68]

On 13 May, the *Chaltham* was ordered from her post guarding the Rufigi to the Dardanelles, the Admiralty having decided to destroy the *Konigsberg* by monitors, their fire being directed by the seaplanes. The Brazilian Navy had ordered three monitors from Vickers before the war and they were almost ready by August 1914, when the admiralty requisitioned them. They were named HMS *Severn, Mersey* and *Humber.* The monitors were of shallow draft and had performed well in the early days of the war bombarding German batteries between Ostende and Nieuport. They proved to be not particularly seaworthy, very slow, 10 knots being their fastest speed and they had a short radius of action. The first two were selected to go to East Africa to attack the *Königsberg.* They were rearmed before sailing for East Africa. Each had a single 6-inch gun fore and aft replacing the original twin 6-inch forward turret, and two single 4.7-inch howitzers aft. The howitzers were relocated to the boat deck, together with four semi-automatic 3-pdr (two per side), a single HA 3-pdr, plus multiple machine guns. The first difficulty to be overcome was how to get the vessels to Africa.

The 15th saw the *Severn* and *Mersey* reach Aden, and they were ordered to Pemba Island, where they arrived on 3 June. In the meantime, four aeroplanes and further seaplanes had been ordered for use in the operation. The only suitable place for an aerodrome was on Mafia Island that had been captured from the Germans on 10 January 1916. The aeroplanes, two Henri Farmans and two Caudrons, arrived via HMS *Laurentic* on 18 June, together with Squadron Commander R Gordon, Flt Lt Blackburn, Flt Sub-Lt Arnold and 13 P.O.s and men. On unloading the aeroplane cases it was found that many of the propellers had warped and were useless, leaving only one spare for one machine and two for another.

Both *Severn* and *Mersey* were battened down and made as watertight as possible. Two tugs accompanied each monitor the crews of which had been taken onboard by the merchant steamer HMS *Trent.* A collier completed the small fleet that made its way to Mafia Island on 2 June after a very slow passage from England. On arrival work commenced on armouring the monitors with old boiler plates over the decks and around the bridge.

No. 122 was flown by Cull with various observers in the interim such as that of 8 June 1915, when he and PC Boggis made a reconnaissance and *sighted smoke of K in same position.* Its last flight was recorded on 14th when Cull and CPO Morrington started on a reconnaissance but the *revs dropped about 20 m from ship – landed found 2 blued cylinders about 5 miles off Ras T Waua.* A motor boat had to take them in tow back to the ship. Engine No. 2963/964 was removed and, as no spare rings nor spare engine were available, the seaplane was condemned on 13 July 1915.[69]

In order to preserve the secrecy of the intended operations, it was decided to rendezvous at MAFIA ISLAND, where an aerodrome was quickly cleared by native labour.[70] The newly arrived aircraft were moved here. One of the Caudrons was lost on its trial flight and a Farman crashed due to engine failure soon after arrival. The other aircraft carried out bombing attacks and reconnaissance for the monitor attacks on the *Königsberg,* success being eventually won on 11 July with the German ship being hit and sunk but not before it had brought down the Farman spotter aircraft being flown by Cull. Looff ordered the ship scuttled and charges in the hull were set off at 2 p.m. and the ship settled into the river mud. Even then the *Königsberg* contributed to the German's East Africa campaign as ten of her 105-mm guns were salvaged and taken inland where they were converted to field artillery. Looff reported to Berlin *SMS Königsberg is destroyed but not conquered.*

The last aeroplane of the unit, a Caudron, was lost after confirming the destruction of the *Königsberg.* In celebration the remains of the Short floatplanes were set on fire on the Mafia Island beach.[71]

With no aircraft the squadron returned to Zanzibar where they were joined up with two Caudrons and three Short 827 seaplanes that had arrived on the *Matiana.* The Admiralty offered the use of the aircraft to the Army and the Caudrons went to Maktau while the Shorts went to Mesopotamia.

Cull wrote that *Looking at the conduct of the enemy with impartial eyes it is only fair to say that the very high degree of gunnery and general fighting efficiency displayed in the action, after so many months of forced inaction in the fever ridden waters of the Rufiji, was worthy of high praise.*[72] Gallehawk also considered that, looking at the *conduct of the crew of the "Koenigsberg," it is only fair to say that a very high degree of gunnery and general fighting efficiency displayed in this action after so many months inaction in the fever-ridden waters of the Rufiji was worthy of high praise.*[73]

Specifications Short S.41			
Serials	**Contract**	**Type**	**Notes**
81 – 82	C.P.35829/13	Admiralty 74 Type	S.64 & S.65.
89 – 90	C.P.34603/13	Admiralty Type 81	3-bay. Cancelled.
119 – 122	C.P.53783/13/X/16128	Admiralty Type 81	S.82 to S.85. 3-bay.
811 – 818	CP02129/14	Type C Folder (Improved Admiralty 74 Type)	S.108 to S.115.

The siege of the *Königsberg* showed the value of aircraft to direct naval fire and to bring back useful information. This was the first time that a capital ship was sunk with the assistance of air power

End Notes

1. 'War Experiences 1914–1918. Wing Cdr John T Babington,' Andover, September 1924. (Third Course). AIR1/2387/228/11/38. Note his name was misspelt. Babbington was transferred to aeroplanes in October 1914, when the Westgate station was temporarily closed. He was destined to remain with aeroplane units for the rest of his time in the RNAS. He took part in the raid on the Zeppelin sheds at Friedrichshafen on 27 November 1914, flying Avro 504 No. 873 (not No. 875 as stated in the article). He went onto fly Handley-Page bombers with No. 3 Wing. In 1918 he was sent to the UK to train future crews. Babington's full report is published in The '14-'18 Journal, 2020 Vol. 1, Journal of the Australian Society of Aero-historians.
2. Babbington records in the Log Book of No. 120 that he went to Westgate from Grain on 15 June 1914, and performed an 'Outer Patrol' the next day from Westgate. TNA AIR1/2056/204/412/120.
3. "Seaplanes at Eastchurch." *The Aeroplane*. 19 March 1914. P.318.
4. In his article Grey said that *on their return to Eastchurch, we were allowed to go over a 100-h.p. machine of similar type, which is illustrated and described elsewhere by Mr. Sayers.*
5. 'The New "Short" Seaplane' by W.H.S. *The Aeroplane*. 19 March 1914. P.316.
6. These scale drawings were used in German publications to illustrate the Short seaplanes.
7. Sturtivant, R & Page, G. *Royal Navy Aircraft Serials & Units*, Air Britain, UK, 1992.
8. "Early British Naval Aviation." This undated paper was written by the then Air Commodore Samson after his retirement. Copy in IWM via G Bruce.
9. The first de Marçay heavier-than-air aircraft was designed with Emile Moonen. The 1911 de Marçay-Moonen monoplane was the first to feature folding wings for transport and storage. Together, de Marçay and Moonen had filed UK Patent 18,661 "Improvements in and relating to Aeroplanes and other Machines" in Oct 1909, describing the folding wings. https://www.secretprojects.co.uk/threads/de-mar%C3%A7ay-aircraft.21456/]
10. Notes from Shorts' Order Book via Gordon Bruce.
11. TNA AIR1/216.
12. Sturtivant & Page name the two as Short Admiralty 74 Type Folder Tractor Biplane Seaplanes.
13. 'Naval and Military Aeronautics', *The Aeroplane*, 24 July 1913. P.94.
14. 'Naval and Military Aeronautics.' *The Aeroplane*, 11 September 1913. P.288.
15. Extract from the Archives of Sir Arthur Longmore. *Manuscript Log of RNAS Calshot 6 May – 24 September 1914*. Copy in FAA Museum.
16. No. 82 was still listed in the December 1914 list of Naval Aircraft, but in the March 1915 list. Letter from J.M. Bruce to G.S. Leslie. 25 March 1986. G.S. Leslie papers.
17. Extract from the Archives of Sir Arthur Longmore. *Manuscript Log of RNAS Calshot 6 May – 24 September 1914*. Copy in FAA Museum.
18. Longmore, A.M. Op Cit. P.36. Longmore retired as Air Chief Marshal Sir Arthur Longmore, GCB, DSO. The photograph in Longmore's book showing the *Short after successful torpedo experiments* is of the Short Improved S.41 No. 74. J.M. Bruce noted that in the text Longmore wrote that he did not carry a passenger and one is shown in the photograph. *All the 160 h.p. Gnome Shorts I have ever seen have had a broad stack exhaust rising from the top of the engine cowl, and No. 74 has no such thing*. No. 74 was not the aircraft that carried out the experiment. Letter to H.F. King, 05 June 1954. RAFM J.M. Bruce Collection. Box 37.
19. Extract from the Archives of Sir Arthur Longmore. *Manuscript Log of RNAS Calshot 6 May – 24 September 1914*. Copy in FAA Museum.
20. Longmore's Log Book confirms that the machine that carried out the first torpedo drop was No. 121. For a discussion of Barnes' claim in his book *Shorts Aircraft since 1900*, (Putnam, UK, 1967), that Gordon Bell made the first torpedo drop, see Bruce, J.M. 'The question of dropping a torpedo.' *Aeroplane Monthly*, March 1987. P. 148. Photocopies of Longmore's Logbook via Gordon Bruce. Longmore's Logbook is held at the National Maritime Museum, Greenwich, London.
21. Handwritten notes from TNA AIR1/186/15/226/3.

Specifications Short Folder Seaplanes			
Source	**1**	**2**	
Type	Nos.81–82	Nos. 81–82	Nos. 119–122
Span upper	56 ft	56 ft	67 ft
Span lower	40 ft		52 ft
Length	39 ft	40 ft	42 ft
Height			
Chord	6 ft		6 ft
Gap	6 ft		6 ft
Stagger	Nil		
Dihedral	Nil		
Airscrew Dia	9 ft 6 in		
Top plane in ft^2			
Lower plane in ft^2			
Wings in ft^2	550	550	690
Empty Wt in lbs	2,000	2,400	3,050
Military load in lbs	80		
Crew in lbs	360		
Fuel & Oil in lbs	600		
Floats[1] in lbs			
Loaded in lbs	3,040	3,100	3,500
Fuel, in gal.	65		
Main tank in gal.			
Gravity tank in gal.			
Oil in gal.	15		
Speed max	78 mph	78 mph	78 mph
Endurance	5 hrs	5 hrs	5 hrs

Sources: 1. J.M. Bruce Data.
2. Goodall & Tagg Data.

RAFM J.M. Bruce Collection, Box 4.

22. O Short to Air Ministry, RE Claim of Admiral Sueter and Lt. Col. Malone in respect of Torpedo Carrying Aircraft. 8 April 1926. TNA TS28/18.
23. Notes by J.M. Bruce from TNA AIR1/186/15/226/3. RAF Museum J.M. Bruce Collection Box 4.
24. The Log Book of No. 120 recorded: *Aug 1914. Chassis altered to carry 1 14' torpedo or 5 100 lb. Bombs.* Work carried out by Messrs. Short Bros at Grain. TNA AIR1/2056/204/412/1/20.
25. Notes from TNA AIR1/185/15/226/2 by Gordon Bruce.
26. Notes from Gordon Bruce, July 2020, and Sturtivant & Page, *Op Cit*.
27. Nos. 119, 120 and 122 were transferred to *Engadine* on 21 November 1914. *Engadine*, *Empress,* and *Riviera* were cross-Channel steamers that were hurriedly converted to seaplane-carriers with *Engadine* and *Riviera* being intended to operate torpedo carrying aircraft in the North Sea.
28. For the story of the Canton-Unnê engines see Cowlan, P. "The Development and use of the Salmson Canton-Unnê Aero Engine 1908-1918." *Cross & Cockade International*, Vol.49, No. 1, P.37 & No. 2 P.114.
29. Handwritten notes from TNA AIR1/186/15/226/3. RAFM J.M. Bruce Collection, Box 4.
30. Handwritten notes from TNA AIR1/186/15/226/3. RAFM J.M. Bruce Collection, Box 4.
31. Bone wrote of his WWI activities in *Cross & Cockade Great Britain*, Vol.9 No. 1. 1978. P.1. He transferred from the submarine service to the RNAS pre-war.
32. 'Combined Air attack on Ostend, Bruges and Zeebrugge by 38 aeroplanes and seaplanes.' Secret. Signed by H.F. Oliver. 5 02.1915. TNA AIR1/301/15/228/48.

33. Bone, R.J. 'A Record of Good Luck.' *Cross & Cockade Great Britain*, Vol.9 No. 1. 1978. P.5.
34. Samson's hand written report. 17.02.1915. TNA AIR1/301/15/228/48.
35. Gerdessen, F & Geldhof, N. *De Internneringen van vliegtuigen tijdens De Groote Oorlog*, Uitgeveri Geromy bv, Netherlands, 2018. P.237. Also, Hobson, C. *Airmen Died in the Great War 1914–1918 – The Roll of Honour of the British and Commonwealth Air Services of the First World War*, JB Hayward & Son, UK, 1995.
36. Rutland, F. J. Air Memoirs, typed manuscript. FAA Museum 1997/234. This is a history of the aircraft carrier written by Rutland.
37. "Navy and Military Aeronautics." *The Aeroplane*, 30 December 1914. P.566. The Admiralty press release stated that German warships lying in Schilling Roads, off Cuxhaven, were attacked by seven naval seaplanes. The pilot's names are given but not the passengers.
38. TNA AIR1/2099/207/23/4/ Air Department, February 1915. See also, Goodall, M.H. "FET Hewlett and the Cuxhaven Raid," *Cross & Cockade Great Britain Journal*, Vol.6 No. 2, 1975, P. 71.
39. Goodall, M.H. "FET Hewlett and the Cuxhaven Raid," *Op Cit*. P. 72.
40. The Log Book of No. 120 records: 25/12/14. 2 hrs 50 min. Flight Lieut Miley. *Carried out reconnaissance flight over Cuxhaven district. Abandoned machine 10 miles N of Norderney Lt. Ho. (Sunk by gunfire)*. AJM. TNA AIR1/2056/204/412/1/20.
41. Rutland, F. J. *Air Memoirs. Op Cit*.
42. According to Hobbs, D, after the submarine submerged to rest on the seabed, the airmen joined the E.11's submariners for Christmas lunch complete with turkey and plumb pudding. (*The Royal Navy's Air Service in the Great War,* Seaforth Publishing, UK, 2017. P.91.
43. The actions of the German Navy's Zeppelins are taken from Robinson, D. *The Zeppelin in Combat*, Schiffer Publication, USA, 1994. PP.69–71. This is still the best account in English of the Zeppelins activities in the war.
44. Rutland, F. J. *Air Memoirs. Op Cit*.
45. The engines were subject to frequent failures. No. 136, the most reliable seaplane, made 26 flights between 17 February and 31 May, but suffered 16 engine failures during this time. Layman, R.D. *Naval Aviation in the First World War,* Caxton, UK, 2002. P.142. John Oliver takes a more jaundice view of the *Ark Royal* and Sqn Cdr R.H. Clark-Hall and Captain Cecil F Kilner in his *The Air War in the Dardanelles* (J Oliver, UK. 2017). He compares Samson's flights from land against the *Ark Royal's*. The comparison is not fair as has been noted, the weather and general conditions of the seaplanes together with the lack of training led to the ship not being able to function as well as land based aeroplanes.
46. 'Mother Ship for Seaplanes.' *Flight*, 14 August 1914, P.853.
47. Layman, R.D. *Op Cit*. P.149.
48. Colonel F.H. Sykes Royal Flying Corps Attached General Headquarters Mediterranean Expeditionary Force. 9.07.1915. TNA AIR1/669/17/122/788. Sykes, not being a Navy man, makes no account of the fact that the whole use of the seaplanes was experimental. No previous work such as this combined operation had taken place.
49. W/T from V.A. 25.02.1915. TNA AIR1/669/17/122/788.
50. To 1st Sea Lord from M Sueter. 25.02.1915. TNA AIR1/669/17/122/788.
51. The German SSM *Weissenburg*, a Brandenburg class battleship, was sold to the Ottoman Empire and renamed *Turgut Reis*, after the famous 16th century Turkish admiral. The ship saw heavy service during the Balkan Wars, primarily providing artillery support to Ottoman ground forces.
52. Jones, H.A. *The War in the Air*, Vol. 2. The Claredon Press, Oxford, UK. P.47.
53. *To this day the Royal Air Force regards this warrior's feats as "brilliant."* Quoted in Chatterton, E.K. *The "Königsberg" Adventure*, Hurst & Blackett Ltd, UK. P.111. For the full story of Cutler and his gallant little flying boat see Jones, H.A. *The War in the Air*, Vol. 3. The Claredon Press, Oxford, UK. PP 2–5.
54. 'Record of Expeditionary Squadron No. 4, Royal Naval Air Service' by J.T. Cull, DSO. Hereafter referred to as Cull's Report. TNA AIR 1/725/106/1.
55. Cull lists the trades of these 18 men as follows: Chief Petty Officers Norrington (ERA); Lacey (Shipwright); Mobsby (from Messrs Sopwith); Sprott (from Messrs Lang); Leading Mechanics Bennett, Boggis (Observers, R.M.LI) Cornelius, Sutcliffe, Air Mechanics Law, Tubbs, Brazier, Tall, Bois, Gregory, Shaw, Quarman, Gant, Williams, Burgess (from Messrs Lang). This listing shows how much Cull respected the efforts of Cutler and Gallehawk who did it all on their own. Ibid.
56. Sopwith bought out the Lang Propeller Co in 1917 and Arthur Lang went to the USA and started the Lang Propeller Co of America Inc.
57. Cull spells the Island as Neororo in his report.
58. Cull's Report, *Op Cit*.
59. Quoted in Contey, F.A. "Siege in the Jungle" *Aviation History,* USA, July 1994, P.27.
60. Log Book Short No. 119. TNA AIR1/2065/204/412/921. This Log is appended in the Log Book of Sopwith No. 921.
61. Boggis was awarded a Distinguished Service Medal for his work in the locating and destruction of the *Königsberg*.
62. 'Record of Expeditionary Squadron No.4, Royal Naval Air Service.' Narrative by Wing Cdr J.T.Cull, RAF. 22 Nov 1914 to Jan 1917. TNA AIR1/725/106/1.
63. Log Book of No. 122 in Log Book of No. 920. TNA

AIR1/2064/204/412/920.

64. 'Record of Expeditionary Squadron No.4, Royal Naval Air Service.' Narrative by Wing Cdr J.T.Cull, RAF. 22 Nov 1914 to Jan 1917. TNA AIR1/725/106/1.
65. Log Book of No. 119. TNA AIR1/2065/204/412/291.
66. Log Book of No. 119. TNA AIR1/2065/204/412/291.
67. Entry 5.05.1915. Log Book of No. 119. TNA AIR1/2065/204/412/857.
68. 'Record of Expeditionary Squadron No.4, Royal Naval Air Service.' Narrative by Wing Cdr J.T.Cull, RAF. 22 Nov 1914 to Jan 1917. TNA AIR1/725/106/1.
69. Log Book No. 122. TNA AIR1/2065/204/412/857.
70. A.C. Bishop recorded that when he arrived at Zanzibar he found *a large naval squadron assembled there, consisting of H.M.S.'s* HYACINTH, CHATHAM, WEYMOUTH, CHALLENGER *and* PYRAMUS *together with attendant colliers and whalers and the two liners* Laurentice *and* Laconia, *which were acting as store ship for the fleet*. All this to protect trade from one German raider. This was a *striking example of how expensive and troublesome a single raider, in bold and daring hands, can prove.* War Experiences of Flt Lt A.C. Bishop. Copy of AH228/11/36. (2nd Course).
71. Cull did excellent work considering the difficulties he operated under, however, for a different view of Cull see 'Extract from Leading Mechanic Edward Robert Wingrove Lincoln's Diary' Mar-Sep 16. Dye, P. "Royal Naval Air Service Operations in German East Africa 1914-1918; Part 3," *Cross & Cockade International*, Vol. 38, No. 1, 2007, P.22.
72. Cull's Report. *Op Cit.*
73. Cull's Report *Op Cit.*

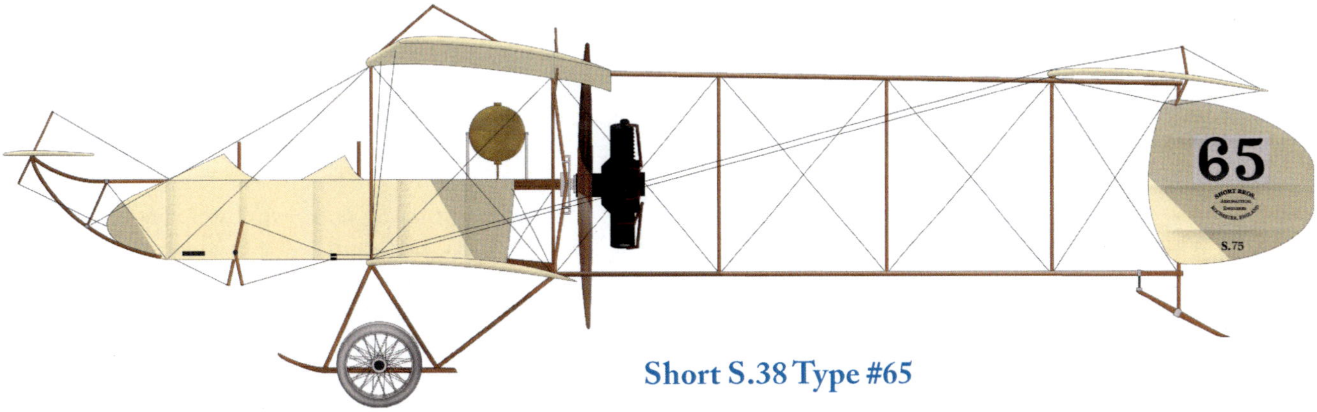

Short S.38 Type #65

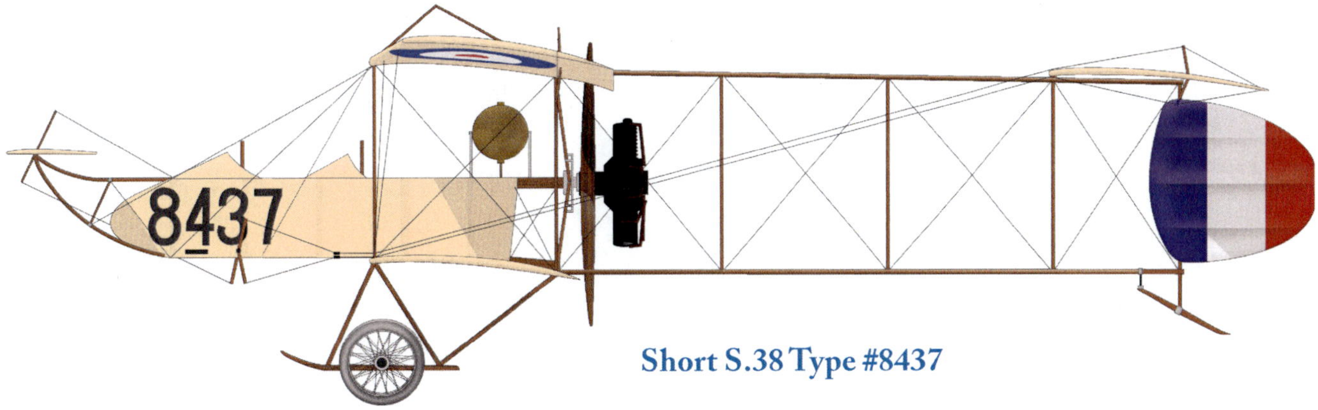

Short S.38 Type #8437

Chapter 6: Short Seaplane Admiralty Type 135

Two developments of the Folder seaplanes were ordered on 12 February 1914, as Short Admiralty 135 Type Folder Tractor Biplane Seaplanes, under Contract No. C.P.30774/14/X.2461. Given the serial Nos. 135 and 136, the tractors were powered by Salmson water-cooled radial engines. No. 135 (S.88) had a nine cylinder 135-hp Canton-Unné, while No. 136 (S.87) had a twin-row fourteen cylinder 200-hp Canton-Unné Type 2.M.7 engine.[1] These French engines were built in the UK by Dudbridge Ironworks of Stroud, Gloustershire.

The two machines were probably identical but no photographs of No. 135 appear to have survived. On the new machines the wings were of two bays with the upper wing extensions braced by angled struts from the lower wing. Ailerons were single acting and were inversely tapered, the taper on their outer portion being less pronounced than that on their inner portion. In order to allow the wings to fold and not foul the bracing cables the outer portions of the ailerons had a long chordwise notch. These notches became a feature of Short seaplanes. The fuselage had a full rounded top decking. The tail unit was generally similar to the earlier Folders.

Being water-cooled the Salmson engines required a radiator. This was a large box-shaped affair mounted on top of the fuselage behind the engine. Although it must have interfered with the pilot's vision it was to become another feature of Short seaplanes, the loss of pilot's view not being such as to cause concern.

Shorts delivered No. 135 to Grain by towing it from Queensboro Pier by a steam boat, arriving at 1600 on 28 August 1914. No.135 was delivered to HMS *Rivieria* in September. Grain reported on the 9th that No. 135 was to go to the ship to test the hoisting arrangement and for test flight. She lasted until 25 December 1914, when she was lost on the Cuxhaven raid. (See Chapter 5).

No.136 was delivered to Grain by Sqn Cdr J.W. Seddon and Mr Fairey on 28 August 1914. With the *Riviera* by 16 September 1914, she also took part in the Cuxhaven raid.

The Dardanelles

HMS *Ark Royal* had been completed in early December 1914, and was working up when she was earmarked by Churchill for the Dardanelles. In a January 1915 memo, Capt Murray Seuter, the director of the Admiralty Air Department, was instructed to hold the *Ark Royal* for service in Egypt.[2]

Ark Royal left Sheerness on 31 January 1915, with Short No. 136, two Wight Improved Navyplanes Nos. 172 and 173; Three Sopwith 807 Type Nos. 807, 808 and 922. No. 136 was reported to be in good order. *Has now had new planes fitted. The most valuable and only rough weather machine in the ship. Has suffered considerably from engine trouble and on stripping down engine a fractured ball race and other damaged parts were found. No spare one aboard. Cabled for, but not yet arrived. These parts now been taken from engine of dismantled No. 173.*[3]

Arriving at Tenedos on 17 February, with the exception of three short expeditions to the Gulf of Xeros for reconnaissance of possible landing places and to spot for ships, one reconnaissance of the Gulf of Adramyti, an attack on a T.B. in Smyrna Harbour and a visit to Mudros, *Ark Royal* remained at Tenedos and Rabbit Islands until the landing on 25 April. Then, being engaged in reconnaissance and spotting on the Dardanelles defences, her seaplanes taking part in attacks on both the outer and inner Forts. The French seaplane carrier *Foudre's* Nieuport monoplane seaplanes also took part in the operation, but, unlike their British counterparts, were not equipped with wireless and were useless for spotting.

The results of the seaplane's work were haphazard. This was due to a variety of factors: - the seaplanes themselves being low powered and unable to operate in the conditions prevailing at an altitude high enough to enable effective spotting to take place. The W/T sets were really still in their prototype phase and the crews had no realistic practice with them before coming to the Dardanelles. The sets could send but not receive. They were large, heavy and only two were originally present. If the machine that they were fitted in was out of commission, they had to be removed and transferred to another.[4]

The monthly report of flights by *Ark Royal's* aircraft noted the following problems with the Shorts amongst the reports on all her seaplanes:

11th April
Short No. 136. Engine test – still failing to give power.

23rd April
Short No. 161. Three short test flights totalling 1hour 14 min. *Machine finally tuned up.*
Short No. 136. After a 22 minute engine test, Lt Dunning with Lt Brown, RNVR, carried out W/T tuning for the Fleet at Mudros.

24th April
Short No. 136 Lt Dunning with Lt Strain were sent to make

Above: This is thought to be Short No. 135 under construction at Short Brothers, Eastchurch.

last reconnaissance of Gabe Tepe before landing. Wind and sea high no flying possible until late when No. 136 got off but came down with a broken inlet valve spring.

25th April

Short No. 161. Flt Cdr Kilner with Lt Parke. Broken oil pump spindle brought the machine down in 10 minutes. No. 161 was replaced by Sopwith No. 1437 piloted by Lt Garnett (solo). This flight lasted three minutes due to pressure trouble.

27th April

Short No. 136. Lt Dunning with Lt Brown. Scouting to locate new batteries but had to return due engine oil on plugs.

No. 136 took part in the Cuxhaven Raid at Christmas. Sent to HMS *Ark Royal* on 31 January 1915, it sailed to the Dardanelles on that ship. Short No. 136 being hoisted aboard with the port float suffering damage to its rear end. 27 April 1915, Kephaol Bay.

Short Seaplane No. 136 returning minus her floats. On 27 April 19115, with a crew of Lt E.H. Dunning and Lt W Parke, on a flight in the Dardanelles to spot the fall of shells from HMS *Triumph* on a Turkish battleship in the Narrows, a float strut and lower wing was shot through by rifle fire, leading to the undercarriage collapsing on alighting in a rough sea. On 23 May, the undercarriage gave way again when flown by the same crew. It was believed that it had struck some wreckage on alighting.

Above: Short No. 136 with a Wight A.1 Navyplane on HMS *Ark Royal*. The Wight used their own method of folding wings.
Below: The sad remains of No. 136. She was transferred to Ispatho Island, Mudros (referred to in Ark Royal's log as H.Q.) in June 1916. The group posing by its remains includes Sylvester Louis Cowlan, seated second left on the transportation dolly. No. 136 was not considered to be successful as a dual control machine and was dismantled for destruction, being written off on 9 June 1916. The machine behind is a Type 166 (sadly unidentified) that is doped in the dark P.C. finish. (via P Cowlan)

30th April

Short No. 161. Pilot Flt Cdr Kilner with Lt Cdr Hornby. Spotted for *Triumph* onto Fort 26 South of Maidos. Also spotted her onto the *Torgud Reis* that came down straights which was shelled but not hit.

Short No. 161. Flt Cdr Kilner with Lt Brown. To scout round Kojasere to locate batteries, but they did not fire when machine was in air. Ball-race in engine found broken when landed.

Short No. 136. Lt Dunning with Lt Parke. To spot for *Triumph* onto Turkish battleship in the Narrows. Ship moved off after a few shots. Proceeded over lines to locate guns. Found one battery. Descended to 2,500 ft to make certain and came under rifle fire. Three longerons shot through and two shots in petrol tank. One undercarriage strut broke on alighting. The machine was out of service for three days as a consequence.[5]

Short No. 136 made 26 flights between 17 February and 31 May, during which it had 16 engine failures, and this was one of the most reliable aircraft Clark-Hall had.[6]

Short 166 Type No. 161 was allocated to the *Ark Royal* and arrived on 19 April and was tested on the 23rd. Capt Robert Hamilton Clark-Hall, Commander of the *Ark Royal*, wrote that the weather conditions were not as might be expected in the Mediterranean, *not at all favourable, and it is for this reason that Short Machines with Canton Unne Engines have been repeatedly asked for.* He noted that these were the best rough weather machines and it was essential that the number of engine types be reduced to a minimum *both on account of the supply of spares and experience gained in practice by Officers and mechanics.* He was particularly concerned that the lack of rough water seaplanes would lead to an *underestimation of the capabilities of seaplanes by Senior Officers.* Four Short Canton-Unnê seaplanes had been asked for by the Commander-in-Chief on 24 February.[7] No. 161 was the only one received to date.

From 25 April to 9 May, *Ark Royal* lay off ANZAC, her seaplanes being engaged in reconnaissance of the surrounding country, spotting for ships firing at targets ashore and at the *Torgud Reis* and other vessels in the Dardanelles and also in flying over the Turkish lines in order to keep down their gunfire as it was noted that when a seaplane was in the air, the carefully concealed field gun batteries would cease fire. The arrival of RNAS No. 3 squadron (later No. 3 Wing) saw landplanes take over the duties of the seaplanes. The seaplanes took great risks in flying at low altitude over the enemy's territory before the aeroplane squadron arrived.[8]

As Layman has pointed out, the wireless sets then in use were barely out of the experimental stage and could only send and not receive. No practice had been undertaken pre-war to develop a system of signals so that the personnel at the Dardanelles had to develop their own method while they were in combat.[9]

The two 225 H.P. Sunbeam Sopwiths are at present on board S.S. "ARAGAZ" in their packing cases. With seven machines on board "ARK ROYAL" there is of course no room for erecting them on board. A tent for use in erecting the machine ashore at Port Mudros had been asked for but not received. The Admiral had been asked for an empty collier or steamer which could be used to carry out repair and erection work. The report concluded with the note that No. 175, a Wight seaplane was being sent out, but the hold would not allow three Wights to be stowed at the same time, *but as one of the three will practically always be requiring overhaul of machine or engine, that one will be kept in steamer or tent ashore till required to relieve another.*[10]

On 9 May hostile submarines made their first appearance and this resulted in the *Ark Royal* was sent to Kephalo Bay, as related in Chapter 5.

During the first five months on station the ship carried one 200-hp Canton-Unné Short tractor seaplane, two Wight pusher seaplanes and three *Daily Mail* Sopwith 807 Seaplanes, together with two Sopwith Tabloid aeroplanes that could fly off the forecastle, but there is no definite record that any ever did. *None of these machines were satisfactory, the Short was old, the Wights frequently failed to get off the water and never climbed to a sufficient height and the Sopwiths climbed*

Ark Royal Reported on 5 May 1915, on the State of its Aircraft:

Serial No.	Engine	Type	Notes
161	200 CU	Short 166	Ready & flying well.
136		Short 135	Brought down by rifle fire on 27th April & damaged in landing. Will not be ready for some days.
172		Wight Improved Navy plane	Awaiting spare ball race from England – cabled for in March.
176			Ready & flying
922	100 Mono Gnomê	Sopwith 807	Ready & flying.
807		Sopwith 807	Undergoing overhaul.
1437		Sopwith Schneider	Ready & flying (in HMS *Doris*).
1438			Ready & flying.

Source: TNA AIR1/661/17/122/641.

so slowly that most of the flight was wasted in getting enough height to go over the enemy. No necessity to use the aeroplanes ever occurred. Spares failed to arrive for long periods leading machines out of action due to lack of replacement parts. The result was that the machines could not be relied on and suffered severely from rifle and machine gun fire. *The condition was not improved by the arrival of two 225 H.P. Sunbeam Sopwiths which hardly did one useful flight and never climbed to a sufficient height.*[11]

Captain Clark-Hall of the *Ark Royal* wrote that No. 136 was the most valuable and only rough-water seaplane he had.

History of Short No. 136

No. 136 had a fair life in service and its Log Book records what was typical of the problems and successes that these early seaplanes encountered.[12]

27 Aug 14. Delivered to Isle Grain by Short Bros.
16 Sep 14. To *Riviera* from Grain.

Riviera
16 Oct 14. Fitted with floats having 40% increase in buoyancy.
25 Oct 14. The floats and chassis were wrecked along with other small damage when the 100-lb TNT bomb was unshipped by the heavy wash of a destroyer and it exploded on the bottom, at a depth of 13 fathoms. Returned to Short Bros for repairs on 31st. Returned to Isle of Mann on 6 Dec 14. After repair trial returned to *Riviera* on 22nd.
31 Jan 15. From *Riviera* to *Ark Royal.*

Ark Royal
05 Feb 15. Fitted with bomb gear to take eight 20-lb Hales bombs.
08 Feb 15. Fitted with new floats and starboard foremost chassis strut.
27 Mar 15. Old lower planes replaced.
27 Apr 15. The chassis was wrecked and lower plane were damaged when the machine landed with the port foremost chassis strut shot through. Chassis rebuilt on board from spares.
02 May 15. Floats and chassis wrecked, prop smashed, leading edge of lower plane damaged when the two foremost chassis struts collapsed as the machine was getting off. These had been fitted after the accident of 27 April and had been made on board from slightly larger thickness tubing in the aviation store, but evidently of softer metal.
23 May 15. Made a good landing and the bottom of the floats collapsed and was torn off. Lower planes, bottom of the fuselage and tail were submerged. The sea was calm and it was thought that the probable cause was striking wreckage while machine was just airborne while getting off. Fitted two new floats, spares for No. 161, and new rear crossbar.
18 July 15. Fitted two additional steel tube struts to fore end of chassis to strengthen same. Fitted kingpost and bracing wires to underside of front cross bar.
29 Jun 15. Forced landing in long swell with no wind. The chassis stood two swells but collapsed on the third. Tops of both floats damaged; prop smashed and the lower planes were submerged.
24 Jul 15. The floats started to split. Probable cause was the joints in the $^1/_8$ in ply with strengthening piece was not strong enough after the duralumin sheet was removed.
18 Aug 15. Bottom of floats damaged and finally collapsed while taxying at 50 knots.
05 Oct 15. Machine stripped, major overhaul. All wiring cleaned, renewed where necessary and repainted. All woodwork scraped and removed where necessary. Planes, fuselage and empennage completely recovered. All controls renewed; new floats fitted but original chassis replaced after complete examination. Two Sopwith pressure tanks fitted on either side of passenger's seat and one behind the seat. Dual control fitted with a complete set of instruments for the passenger.

Aluminium sheeting placed on the fuselage sides as far back as the pilot's seat. The original streamlining was replaced and covered with fabric. This work was completed on 21 December.

09 June 16. Machine completely dismantled for destruction and written off charge due to age.

In April 1915, *Ark Royal* received its first Short 166 Type seaplane, more arriving in July.

End Notes:

1. The engines were of Swiss design and were referred to as Canton-Unné engines from the names of the joint designers of the Salmson engines.
2. Layman, R.D. *Naval Aviation in the First World War*, Caxton Editions, UK, 2002. P.141.
3. Commanding Officer *Ark Royal*, Commodore Robert H Clark-Hall, to Director Air Dept, 11.04.1915. TNA AIR1/661/17/122/641. Clark-Hall had been the Sqn Cdr for the aircraft on HMS *Hermes*. *Ark Royal* was his first solo command.
4. Layman, R.D. notes that the only such test that he could find was a 1907 experiment by the battleship *Revenge* with a man-lifting kite. The last time that the British Fleet had carried out large scale shore bombardment was the bombardment of Alexandria in 1882. *Op Cit.* P.143.
5. H.M.S. "*Ark Royal*" – Monthly Return of Flights. April 1915. TNA AIR1/661/17/122/641.
6. Layman, R.D. *Op Cit.* P.142.
7. Commanding Officer *Ark Royal* to Director of Air Dept, 11.04.1915. TNA AIR1/661/17/122/641.
8. Vice Admiral J.M. de Robeck to Admiralty. "*Ark Royal*" – *Flying Report for Month of January 1916 and Notes on the year's Work 1915–1916.* 11 Feb 1916. AIR1/649. Admiral of the Fleet Sir John Michael de Robeck, GCB, GCMG, GCVO, was born on 10 June 1862, in

Specifications No.135	
Source	1
Span upper	52 ft
Span lower	
Length	37 ft
Height	
Chord	
Gap	
Dihedral	
Wing Area	530 ft^2
Empty Weight	2,700
Loaded Weight	3,400
Speed, max	65 mph
Endurance	4.5 hrs
Engine	135-hp Salmson

Ireland. In the early years of the 20th century, he served as Admiral of Patrols, commanding four flotillas of destroyers. De Robeck commanded the allied naval force in the Dardanelles during the War. He took control of the campaign to force the straits when Admiral Sackville Carden fell sick. The second attempt, launched on 18 March 1915, was nearly successful, as the Turkish land-based artillery almost ran out of ammunition: however, mines laid in the straits led to the loss of three allied battleships and he withdrew. This led to the subsequent failed ground campaign. He went on to become Commander of the 3rd Battle Squadron of the Grand Fleet and then Commander of the 2nd Battle Squadron of the Grand Fleet. After the war de Robeck became Commander-in-Chief, Mediterranean Fleet and British High Commissioner to Turkey, and then Commander-in-Chief, Atlantic Fleet. He died in January 1928.

9. Layman, R.D. *Op Cit.* P.143.
10. Cdr Clarke Hall, Commanding *Ark Royal* to Director of Air Dept, Admiralty. 05 May 1915. TNA AIR1/661/17/122/641.
11. *NOTE on The Years Work of H.M.S. "Ark Royal" 1915–1916.* TNA AIR1/649/17/122/48.
12. No. 136 Log Book. TNA AIR1/2056/204/412/136.

Specifications No. 136			
Source	2	3	1
Span upper	54 ft	54 ft 6 in	54 ft 6 in
Span lower	40 ft	40 ft	
Length	39 ft	39 ft	40 ft
Height	11 ft 6 in	12 ft 6 in	
Chord		6 ft	
Gap		6 ft	
Dihedral	Nil	0°	
Wing Area	544 ft^2		570 ft^2
Empty Wt	2,700 lbs		3,000 lbs
Loaded Wt		3,700 lbs	3,700 lbs
Speed	60.2 kts @ 500 ft	60–65 mph	72 max
Climb to 3,000 ft	37 min 40		
Fuel, gallons	75		
Oil, gallons	4		
Endurance		4 hrs	4 hrs
Engine	200-hp Canton-Unné Type 2.M.7 No. 220		200-hp Salmson

Source: 1. Barnes, C.H. *Shorts Aircraft since 1900*, Putnam, UK, 1967.
2. No. 136 Log Book. TNA AIR1/2056/204/412/136.
3. J.M. Bruce data.

Chapter 7: Short 166 Type (Type C Tractor Seaplane)

Above: No. 161, the prototype for the type, with the Union flag added, in front of the Eastchurch sheds. The caption stated it was with HMS *Ark Royal* on 2 October 1915. The spreader bars on the floats are the curved torpedo carrying variety. The Shorts logo on the rudder reads 'Short.Bros – London & Sheppey' around the following text in the middle of the oval created by the logo – 'Aeronautical Engineers'; and at the very bottom of the rudder is the Short c/n - 'S 90'.

The Short 166 Type was developed from the Short 135. This new seaplane was designed to carry a torpedo[1] and was powered by a 200-hp Salmson Canton Unné 2.M.7 fourteen-cylinder twin-row radial water-cooled engine. Initially known as the Short Type C it was later given Admiralty identification as the Short 166 Type, apparently after the serial number of the last machine of the first batch. The Short 166 Type was designed to fit the hangar of the aircraft carrier HMS *Ark Royal*.

The first machine, No. 161, was a typical Shorts two-bay seaplane with the upper wing overhang being braced by cables to kingposts mounted above the outermost interplane struts, unlike th-e strut supported overhang on the Short 135. For a small increase in drag, the problem of damaging the lift struts when folding the wings was overcome. The fuselage was a conventional wire braced wooden structure with rounded top decking. A large curved fin, that was to become a feature of Shorts products, helped to support the balanced comma shaped rudder. The ailerons ended at the outer interplane struts and as they had no return cable or balance springs, and hung down when the machine was not under power. On this machine the mainplanes could be folded and by not having interconnected ailerons, the difficulty of connecting controls after folding was eliminated.

The twin main floats were of ash with three-ply covering and were sprung. They were supplemented by a tail float fitted with a water rudder. The steel spacer bars to the floats were arched in the middle to accommodate a torpedo. Wing tip floats were located under the lower wing at the outer interplane struts. No. 161 was fitted with a Rouzet W/T but no receiver. It used a single trailing wire aerial and was powered by an engine driven dynamo.

Photographs of No. 161 (S.90) show it fitted with the arched cross bars on the main floats indicating its torpedo carrying design.[2] The late delivery of these machines was due to the delays in the delivery of the 200-hp Salmson engines.[3] A second small batch of eight seaplanes (811 to 818) was ordered as Short Improved Admiralty 74 Type. According to Sturtivant and Page, these were Short Type C Folders and listed with the Canton-Unné powered models as 166 Type. They are covered in Chapter 5 as Short Folders. Machines from this batch took part in the Cuxhaven raid.

HMS *Ark Royal*

As described in Chapter 6, the first of the new 200-hp Short seaplanes for the *Ark Royal* arrived in April 1915, and *proved satisfactory in all respects.* These latest Shorts did not have to have the engine continually run at full power, thereby improving the engine life and flying statistics for the

Above: No. 161 (S.90), the first of the Short Type C tractor seaplanes that later became the 166 Type. The location appears to be the Aegean.

Above: View of the bow of HMS *Ark Royal* with No. 161 behind the two cranes. A Wight seaplane is in the background.

Above: No.161 taxying at Mudros. The Log of No. 161 is reproduced in the main text.

Above: No. 162 taxying and lifting off from the harbour at Mudros shows that the upper surface of the wings as well as the fin were overpainted in an attempt to camouflage the aircraft. Red ring RNAS national markings on the wings, Union flag to the fuselage and a stylized '162' overlaid on the rudder stripes. No. 162 was delivered to HMS *Ark Royal* on 14 July 1915.

Left: Camouflage is displayed by No. 162 taxying on a calm Aegean Sea.

Below: Flt Lt Norman Sholto Douglas (left) and Flt Cdr Cecil F Kilner with No. 162 on the beach in Kephalo Bay, at the Dardanelles, before its rudder stripes were added. Note the canvas engine and airscrew covers.

Below: No. 163, photographed in Kephalo Bay, shows the camouflage pattern was applied to the fuselage as well as the wings, etc. There was no standard to the application of the dope, each machine bearing a unique pattern. RNAS red ring markings are applied to the wings.

ship. At the end of July other 200-hp Shorts arrived. These were Short 166 Type and from that time the complement of the ship comprised only 200-hp Shorts and Sopwith Schneider Cup seaplanes, *one of which was invariably found capable of carrying out the work entrusted to Ark Royal.* The first Schneider Cup Sopwiths had arrived in April and this type of machine proved useful in reconnaissance where detailed information was not required. They were subsequently used with a certain degree of success for solo spotting and for photography.[4] The senior flying officer of *Ark Royal,* Capt Cecil F Kilner, took a liking to No. 161 and flew it every day for the rest of the month after its arrival on 23 April 1915. The following month he alone flew it on 15 separate occasions.[5]

Flt Cdr G.R. Bromet of the *Ark Royal* recalled that he gave up his Wright Type A.I Improved Navyplanes and *was flying a Sopwith tractor fitted with a 225 h.p. Sunbeam. This type compared most unfavourably with the Short tractors* that the *Ark Royal* had at this time. He noted that when Ben-my-Chree arrived, their Sunbeam Short 184 Types *were not thought much of at the time and I think this was principally due to the engine which was most unreliable.*[6] This is a backhanded compliment for the Canton-Unné Shorts, but a compliment nevertheless.

On 31 October 1915, the ship sailed for Port Iero where several days were spent reconnoitring Smyrna and the coastal towns after which the ship was sent to Salonkia, where apart from one short expedition to direct gunfire at Dedeagatch, she remained, originally patrolling for submarines and afterwards keeping the Grecian Defences under observation and photographing the country for the purpose of map-making.

In addition to the work from the ship, seaplanes were lent on several occasions to the cruisers *Doris* and *Minerva,* and the monitors *Roberts* and *Raglan,* for spotting and reconnaissance work in the Gulf of Xeros, Asiatic Coast, East of Rabbit Islands and in the Smyrna district.[7]

For example, in September 1915, Short 166 No. 164 was transferred to the monitor *Roberts* at the Dardanelles. She spotted for the monitor's guns that bombarded Turkish gun batteries. While on *Roberts,* No. 164 had been secured on the ship's forecastle deck, abaft the funnel subjected to the concussion of the ship's big guns. The seaplane had to be hoisted over the side before action to avoid blast damage from the ship's guns whether the machine was required for spotting or not.[8] On 14 January 1916, No. 164 was recovered by *Ark Royal* from the *Roberts.*

This seaplane had been with the "Roberts" at Rabbit Islands since 9th September 1915 and had only made 10 flights of 11 hours in all.

During that time French aeroplanes did most of the spotting for monitor stationed at Rabbit Islands, No.164 being retained on board the "Roberts" as a 'stand-by'. In being transferred from "Roberts" for conveyance to "Ark Royal" at SALONIKA, the seaplane was dropped from a crane and suffered serious damage, necessitation complete overhaul which is being undertaken aboard.[9]

Summary of Flights, 1915

Class of Service	No. Flights	No. that Were Successful
Reconnaissance	180	134
Spotting	113	78
Photography	22	20
Attack	17	10
Patrol	36	25
Transit	35	31
Experimental	13	10
Trial	81	81
Total	**497**	**389**

Notes: On many occasions two services were accomplished on the same flight (e.g. Spotting & Reconnaissance); in these cases the primary purpose of the flight was taken in determining the class in which it should be counted.
"Patrol" includes both submarine patrol and patrol over the enemy's lines to keep down gun fire.
"Transit" flights included flights from ship to land and vice versa when this was the sole purpose of the flight.
Source: Note on 'The Years Work of H.M.S. "*Ark Royal.*" 1915–1916.' TNA AIR1/649/17/122/418.

At the end of the year *Ark Royal* had five Short 166 seaplanes[10] and it was reported that the Short seaplanes had done considerable work and were now being given a complete overhaul in rotation.[11]

In all, 497 flights were made during the year aggregating 438 hours 48 minutes – 267 of these flights successfully accomplished the war service duties required of them.

The above table sets out the particular class and number in each category which were successful.

The "Flying Report" for January 1916, noted that during the month there had been ten flying days with 11 service flights, aggregating 14 hours 35 minutes flight time. The weather conditions of fog, rain or high winds on the remaining days prevented flights taking place, or the authorities did not require assistance.

In the early part of the month most of the flying consisted in keeping the Greek defences of Salonika under observation, photographing them and photographing the surrounding country for the purpose of map making.

In the operations off DEDEAGTCH on the 18th January Ark Royal's seaplanes were detailed to spot for HMS *"Havelock", her objective being BODOMA Railway Junction, 7000 yards inland behind a hill. One flight of 2 hours 50 mins was sufficient to spot the 26 rounds of 14" allotted to the "Havelock", whose shooting was excellent, damaging the water tower and station buildings and tearing up the railway lines. Photographs of*

Junction and other features of military importance were taken. [12]

Combined operations on the 28th during the occupation of the Kara Point defences saw the *Ark Royal's* seaplanes continually in the air, keeping W/T communication between the ship and other vessels, reporting on the process of events, and standing by to immediately spot for the pre-dreadnought *Prince George* onto any recalcitrant fort. At the end of two hours a second seaplane took over and remained until all the forts were occupied.[13]

Fortunately, the Aircraft Log Book for No.161 has survived and the following account of its activities would be similar to all the aircraft operating from *Ark Royal* during this period.

Allocated to the *Ark Royal*, the prototype, No. 161, was crated and shipped via the SS *Carrigan Head*, arriving on 19 April 1915. She was erected on the *Ark Royal* and ready for flight by a.m. on the 22nd, but owing to a slight engine defect did not actually fly until a.m. on the 23rd, with Capt C.F. Kilner at the controls and Lt W Park, RNVR, as observer with only 70 gallons of petrol. The machine proved to be very tail heavy. This was overcome by giving her increased lift on the tail and fixing a Sandow on the Cloche. On it second flight the adjustments to the tail corrected this and it made 2,000 feet in 10 minutes. A reconnaissance flight on the 25th was cancelled after 10 minutes when the spindle in the oil pump sheared. The same day the pair made a reconnaissance of the Gaba Tepe headland that overlooks the northern Aegean Sea. The *Machine flew well. Barograph broke at 6500 ft. Climbed 3000 ft in 17 minutes.* The Short Self Registering Maximum Height Recorder No.311 collapsed on reaching a height of 6,500 ft. It was replaced by instrument No.199.

Further reconnaissance flights continued for the remainder of April with one on the 27th of Gaba Tepe *with main objective of spotting batteries controlling fire onto them. Nothing seen. Height 6400 ft.* Spotting occurred for *Triumph, Prince of Wales* and *Queen Elizabeth.* On 1 May, Kilner and Park were to *Observe if possible batteries firing, & also to keep down gun fire during infantry attack. Height 6400 ft.*

On 13 May, both starboard after chassis struts and one port after chassis strut fractured on take-off. She struck the crest of a long roller after being thrown into the air by the previous wave before flying speed was reached. Kilner and Park were again the crew. They seem to have been the pair to strike all the bad luck with No. 161. June 20th saw engine No. 221 removed and replaced by No. 265. The 5 July saw the floats damaged when the elevator stuck. August saw the Shorts airscrew replaced by an Integral design (Dia: 280; Pitch 190), that gave 1,350 revs at full throttle. The Shorts airscrew was replaced on the 11th. At the same time new main planes, rudder and elevator were fitted. On 16 August, the pair were again the crew when the bottoms of both floats tore off on landing after spotting for *Theusus. Landing was a trifle fast* and the floats had suffered from the heat and the butt ends tore away. The machine was salved and repaired, and back in the air on 1 September 1915.[14]

November saw a fitting to hold two 20-lb bombs installed inside the fuselage. In April 1916, a small hole (10 ins x 4 ins) was cut in the deck under the rudder bar to allow the pilot to see objects underneath the machine - *Doesn't affect strength at all.* The last entry under "Alterations Made" for *Ark Royal* dated 16 June 1916, states: *To be fitted with land chassis.* On 21st she was transferred to No. 2 Wing, Mundros.

On the 16 June, she had been described as: *Fabric very soft and flabby.* No.161 served at all tasks allotted to her, but was not in a good condition when transferred to No. 2 Wing, Mudros, on 21st. No.161's Log ends here with a total flight time of 77 hours 45 minutes; however, one test flight has no figure attached. No further material on its subsequent history has been discovered to date.[15]

Sqn Ldr Conway Walter Heath Pulford, OBE, AFC, wrote of his experiences with *Ark Royal.* He had left the Fleet to learn to fly at Hendon on 14 December 1914. He had not wanted to leave the Grand Fleet as he had trained for nine years and was hoping for the *big Fleet action which we all hoped and prayed for.* He passed out of the CFS the following March. He was annoyed at being taught Army matters – *I could not see what on earth a Naval officer wanted to know this stuff for, forgetting that in all probability we might at any time be working with the Army. However, if that contingency was likely, why call us the Naval Air Service.*

From the CFS he was sent to Hendon as second in command. Here he had to instruct pupils as well as repel Zeppelins. *Our machines consisted of Boxkites, Maurice Farmans, Sopwith Pushers, two-seat scouts, one Deperdussin monoplane, and two B.E.2c's….I only went up twice (on an anti-Zeppelin flight); once I got down alright, on the other occasion I hit a tree some 60 feet high and completely wrote off the machine.*[16] This occurred in B.E.2c No. 966 on 1 June 1915.[17]

I left Chingford in June 1915, to join H.M.S. Ark Royal, seaplane carrier in the Aegean. I shall never forget on asking the Naval Intelligence Officer on board Europa at Mudros where the Ark Royal was. His reply that "he really didn't know, but he thought she was at Imbros" filled me with contempt for his ability. Such a thing would never have occurred in the Grand Fleet.

Our work in the Ark Royal consisted of spotting for ship's guns firing at targets on the Peninsula. Reconnaissance of the strait of the Gulf of Xeros, submarine patrolling and photography, with occasional bomb-dropping. We worked from a shore base, with our machines out in the open, dragged up on a sandy beach when not in use. They stood up extremely well to what I think were exceedingly trying weather conditions for machines.

Our photographic equipment consisted of two aeroplane cameras R.F.C. pattern, and one German camera, which I think had been captured early in the war. Good oblique photographs were taken, but vertical photographs, though taken, had to be taken from single-seaters as two-seaters could not get the height…

We had no guns to defend ourselves with till October 1915 when we got two Lewis guns. Our wireless equipment consisted of "Sterling" transmitters which worked quite well. Intercommunication from the ship to machines was established

Right & Belows: No. 165 in the original finish of these 166 Type Short seaplanes. The serial is carried on a white field.

Below: The machine that gave its serial to the type, No. 166 on HMS *Ark Royal's* crane. The forward fuselage and floats appear to be white. The vertical box type radiator was a feature of Short seaplanes. The seaman on the boom against the ship's side has a pole he will use to keep the machine from striking the ship. No. 166 also served time on HMS *Roberts* and HMS *Empress*. For a short period, No. 166 was converted to a landplane.

Left & Below: Bomb rack on a Short 166 Type in the hangar of the seaplane carrier HMS *Vindex*, is from the second batch of 20. Built at the Westland Aircraft Works in Yeovil, Somerset, they were referred to, within the company, as the Short 200 Canton-Unné.

Right: Launching a Short 166 Type, either No. 165 or 166. from A Squadron, No. 2 Wing, RNAS base at Stavros. Attached to the side of the fuselage is a bomb sight. Running behind the sight is a vertical rod topped with a U-shaped crutch. This is thought to be a steadying support for the arm of the observer when operating a camera. The bomb load consists of three 100/112-lb bombs slung between the arched float supports and four 10/20-lb Hale type bombs beneath the fuselage. It has what appears to be a bomb-sight on the fuselage side.

Left: No. 164 from HMS *Ark Royal* is shown aboard the monitor HMS *Roberts* while anchored off Rabbit Island. Note the boom to move the seaplane on and off the ship.

Right: The float support struts appear to have sheared off this Short 166.

Above: Short 166 Type in flight with the red/white/blue cockade national markings in unusual proportions to the wings, but the Union flag to the fuselage.

Above: Unidentified camouflaged Short 166 at Stavros in February–March 1916 before the slipway was built. The monitor in the background is of the class of small monitors with a single 9.2-inch gun forward.

Above: The same location but in March 1916, with the new slipway in the foreground. The unusual proportioned wing cockades are well displayed.

by searchlight. At first this method proved slow, on account of the failure of the searchlight personnel to keep trained on the machine, but after a little practice became quite efficient. At first the code used was Over, Short, Right, or Left, but after a while the "clock code" was adopted.

Owing to the difficult experienced by new observers in picking out the correct target, we made plans to design a miniature "Peninsula" on the ground, putting in the various battery positions and teaching observers to recognise them. However, it came to nothing as very shortly afterwards the Ark Royal left Imbros for Salonica.

Here our duties were:
(1) To photograph the Greek defences which each day were being put up with the obvious intention of being used against the Allies when occasion arose.
(2) Submarine patrols in the Gulf of Salonica.
(3) Photographic survey of Salonica Harbour which apparently not been surveyed since 1840 or thereabouts. …

It was a most bizarre situation at Salonica when we first arrived. The German, Austrian, Bulgarian and Turkish consuls all continued to function until about a month after we had landed, when Sarrail arrested them all. The town was brilliantly lit up at night and this probably accounted for the good shooting made by the Zeppelin.[18] *…*

In March 1916 our machines were flown to Stavros, a position on the right flank, where their duties were to make an artillery map for the artillery, and submarine patrols. Personally I had to accompany the Ark Royal *back to Mudros, where my duties were to be in charge of the Flying and Repair work.*

It was here that I picked up my first experience of torpedo dropping, my machine being a 200 H.P. Canton Unne Short seaplane; torpedo carried being a 14" Mark X, weight 750 lb. The origin of my taking up this work was due to Wing Captain Scarlett, who sent for me and informed me that I would have to torpedo the Austrian dreadnought "Radetsky" *about which there was considerable uneasiness. It was thought that she might attempt to make a junction with the* "Goeben" *and this combination would "mop up" anything they came across in the Mediterranean.*[19]

In my report after trails had been carried out I reported that the torpedo was too slow, and a faster torpedo was necessary, but that it was an easy enough thing to hit a ship at anchor.

I made certain expeditions with ships to bombard various points near Smyrna, my machine being used to spot the fall of shot. I always thought it very stupid to take one machine, because if the machine broke down the ship must go back to the base and embark another. On one occasion this happened to me, due to the fact that the machine's undercarriage gave way on landing, as a direct result of being dropped eight feet into the water from a height of eight feet, due to the topping lift of the "hoisting in" derrick breaking when the machine was being "hoisted in" beforehand.[20]

On Wednesday, 22 March 1916, H.R. Hadfield recorded in his diary that No. 161 was being fitted for a torpedo for a trial drop and run. The following day *Pulford out with 163, just gets off water when torpedo drops & splits his propeller & also breaks his tail float & has to* (be) *towed in by motor boat.* On the 29th Pulford is successful. *Torpedo in No. 163 successfully launched it, some sight that snap.*[21]

Pulford was invalided Home after *a nose-dive into the sea on a single-seater*. This was in Sopwith Schneider No. 1577 on 8 May 1916. He had previously had a crash in a Schneider on 29 September 1915, when he was forced to land No. 1437 downwind and overturned.[22]

In February 1916, four of *Ark Royal's* Short 166 seaplanes, Nos. 162, 164, 165 and 166, were flown to a shore base that was established near the picturesque village of Stavros, situated some 40 miles east of Salonika in the Gulf of Orfano. On the shore a Bessonneau tent was erected. The seaplanes appear to have been employed in conducting a photographic survey of the area. Nos. 161 and 163 remained with the *Ark Royal* during this period.

On 3 May the Stavros machines took part in an offensive against Bulgarian lines of communication. The weather prevented aircraft from more distant locations participating in the raid. The seaplanes from Stavros and HMS *Empress* only taking part. They bombed a bridge building depot at the Mesta River, some barracks at Ferejik, and a number of barges at what was believed to be at a submarine base at Fener Point.[18] It appears that around this time the Shorts were fitted with a mounting for a Lewis gun for the observer.[23]

The survey work had been completed by the middle of 1916. Reports that the Bulgarians were constructing a submarine base on their coast led Short 166 seaplanes from Stavros returning to the *Ark Royal.* A new station was established on the north-west of the island of Thasos. A new airfield had also been constructed on the island. Anti-submarine patrols were to be the main task of the seaplanes.

According to Sturtivant & Page, Nos. 161, 162, 163 and 166 were converted to landplanes for use from the Thasos airfield. Cowlan notes that the changed circumstances of the war probably led to the need for seaplanes spotting for the ships being required less, but the need for reconnaissance aircraft increased. This may have been the reason that the conversion was gone ahead with in order to utilise the machines to advantage. *A wheeled undercarriage would allow* Ark Royal's *aircraft to operate with impunity from the airfields on both Imbros and Thasos.*[24] There is photographic evidence that No. 163 was so modified.

Westland-Built Type 166[25]

In response to a *recent telegraphic request* from *Ark Royal*, it was proposed in March 1916 that the Canton Unné Short floatplanes on order from Petters Ltd (Westland) should be fitted with torpedo slinging gear. *The firm already have in their possession a dummy torpedo and necessary slinging gear drawings.*[19] Westland was duly informed that torpedo suspension gear would be required to be provided and fitted in all the machines on contract.[26]

The Short 166 was the second type of Shorts aircraft to be built by Westland in 1916, the first being the Short 184. As with other Shorts aircraft full drawings were not available and considerable work had to be undertaken between the Westland and Shorts technical staff before full working plans were available. According to Harold Penrose, Arthur Davenport was placed in charge of work of processing the drawings issued to Westland by Shorts. These machines were ordered without the torpedo gear and this required redesign by Bruce and Davenport, the Westland design team, to eliminate the provision to carry the torpedo and installing a provision to carry bombs. The order to change back to the original torpedo carrying version must have caused consternation in the Westland design office.[27]

Despite this, the first Westland Short 166, No. 9751, was delivered to Hamble on 1 July 1916, without the arched spreader-bars. However, photographs exist showing Westland built 166 Types with the torpedo carrying gear. In October 1916, it was reported that the last two seaplanes of the Short Type are now being erected and waiting for their 200-hp Canton-Unné engines.[28]

Only 20 Short 166 seaplanes were constructed as the Westland works was being enlarged and the company concentrated on landplanes for the remainder of the war. The Westland Short 166 seaplanes saw action as patrol aircraft in the UK. HMS *Vindex* operated six at various times. Their routine, boring, arduous anti-submarine patrols were necessary to protect Allied shipping. *Vindex's* aircraft operated off the east coast of England.[29]

On 22 October 1916, No. 9757 piloted by Flt Cdr H.F. Towler, with CPO A Blackwell, and No. 9760 piloted by Flt Lt F.N. Halstead with Lt Erskine Childers as observer, left *Vindex* to carry out a reconnaissance of the Schillig Roads. It was planned to use the new high speed Coastal Motor Boats (CMB) to make a raid against the German ships at anchor. Childers was probably chosen for his pre-war experience in sailing these waters.

Immediately her aircraft were launched and airborne, *Vindex* turned and ran at full speed back to the Humber. After about half-an-hour No. 9757's engine misfired badly but it cleared itself and they proceeded until they hit a thick fog bank. They alighted and decided to try again but failed to find any gaps in the fog and so returned to the rendezvous point where they were picked up by HMS *Carysfort*. Although the raid never took place, the reconnaissance was held to have been a success.

No. 9760 also hit the fog bank and decided to return. Once above the fog they sighted Heligoland and set towards it. They were able to take photographs of the ships in the harbour before two enemy seaplanes were seen climbing towards them, so they turned west and headed for the rendezvous. The hostile seaplanes followed them until after 40 minutes the Squadron was sighted and they broke off the

The flying machine resting in its normal position on the battleship's deck.

Getting ready for a flight; the hydroplane is hoisted from the ship's deck.

Above & Below: This set of photographs was published in the *New York Times Mid-Week Pictorial* for 18 May 1916, showing British floatplanes in Salonika. The ship is not a battleship but the seaplane carrier HMS *Ark Royal*, and the machine featured is No. 166.

It is dropped into the water; the bay and city of Saloniki show in the background.

Off for a scouting cruise; the hydroplane rises after a short spin through the water.

Below: No. 164 being hauled onboard the monitor HMS *Roberts*. The style of louvres on the engine panels are an identification feature for the type. The ship in the background is probably the small monitor HMS *M20*. (Bibliotheque Nationale de France. N6946631)

Operation of the RNAS from Carriers

HMS *Vindex* was one of the earliest Carriers commissioned, and Sqn Ldr E.B. Beaumau, RAF, recorded post-war that the *chief difficulties arose from the relation between the Naval Officers and the R.N.A.S. Officer over questions of flying and ship's routine.*

The Captain of the ship, although he had never seen any aircraft before, asserted that everyone and anything in the ship came under his direct command, and he often demanded patrols when conditions were not suitable.

The position of Senior Flying Officer was, therefore, one of extreme difficulty, and required considerable tact and judgement, in order that the Captain might have his wishes carried out, and yet that the flying personnel might not be required to perform duties beyond his powers.

Beauman noted that the situation had not been fully resolved after the war, and he proposed that in future captains of aircraft Carriers should be attached to an Air Force Unit for some months before taking command of such a ship.

Source: TNA AIR1/2386/228/11/5.

chase. Halstead alighted and they were picked up by HMS *Conquest.*

In June 1917, No. 9751 was the last Type 166 to be operated by *Vindex.*

Nos. 9752, 9755, and 9759 were shipped to Mudros. Nos. 9754, 9758 and 9762 to 9764 were sent to the Aegean where they served at one time with "A" Squadron, RNAS. They took over the duties that the first six Type 166 seaplanes had performed.

Westland built machines probably had their serial marked under the tailplane as did No. 9751, the first built by Westlands. This makes the identification of Westland built machines almost impossible. One identification feature that the Westland Short 166 kept from the Short 184 seaplanes that the firm had constructed prior to the 166 Type, was the three wide openings along the bottom longeron for foot access and only two, not three, foot access points under the cockpits.

Sqn Ldr E.B. Beauman served at Calshot in 1915 while under instruction and participating in anti-submarine patrols. *At this time, when an observer was required for a seaplane, no particular officer and man was detailed, but any mechanic standing near the machine was told off to go.*

The results of this were always bad, as Anti-Submarine patrolling above all things, required an Observer with a long and technical training, before he is of any real value for this work.[30]

Norman Woodhead flew No. 9756 while at Calshot. This seaplane had been with HMS *Vindex* when it was damaged in an accident while taking-off on 19 August 1916. Sent to Westlands for repair the following month, it was at Calshot by 13 November 1916. *He found this Cantonné (sic) engined machine lighter on the controls than the 184.*[31]

Discussion of the training machines at Calshot on 7 August 1917, recorded that 102 machines were at Calshot and its sub-station of which 35 were ready, 63 out of action, 3 erecting and 1 awaiting deletion. Of the 63 out of action, 32 were laid up for structural repairs and 31 for engine repairs. The low proportion of machines ready was accounted for by firstly, *the great age of a large number of school tractors, and secondly by the large number of different types of machines, there being 15 types in all, which of course means considerable delay in obtaining spares even for the least defect; as an instance the 200 Canton Unne Short, for which it is very difficult to get spares.*[32]

The Canton-Unné Shorts had served well and deserve to be remembered.

Endnotes

1. They are usually shown as bombers, however in the Shorts' Order Book the entry for No. 161 (S.590) is shown as a torpedo aircraft. Nos. 162 and 164 were also fitted with the arched torpedo spreader bars, but were dispatched to the Dardanelles.
2. Photostat of the photograph album No.3 prepared for Oswald Short by E.E. Galloway, 1919. The caption reads: *The Short 200 h.p. (Canton-Unné) Seaplane was designed about the end of 1913, the first machine being produced in early 1914. It carried a torpedo slung between the main floats, and was one of the earliest machines to be fitted with the Short patent folding wings, the 160 h.p. Gnome being actually the first to be so fitted.* Via Gordon Bruce.
3. Harold Penrose appears to have confused the type with the earlier 135 Type and records them as being fitted with 135-hp Canton-Unné engines until the 200-hp version was available, rather than two different types.

Serial Allocation

Manufacturer	Serial Nos.	Contract No.	Notes
Short Bros	161 – 166	CP37385/14	200-hp Canton-Unné. Type C Tractor seaplanes. S.90 – S.95. Ordered as Torpedo aircraft.
Short Bros.	811 - 818	CP02129/14	100-hp Gnome. Improved Type 74. S.108 to S.115.
Westland	9751 - 9770	CP105566/16	200-hp Canton-Unné 2.M.7. Short S.90 Admiralty 166 Type.

Above: No. 164 on the water. (Bibliotheque Nationale de France. N6946632)

Above & Above Right: The wing-walk is apparent at the starboard lower wing root on Short 166 No. 9760.

Right: No. 9760 in a spot of bother. Note the crewman clinging to the fuselage.

This Page: No. 9751, the first Westland Short 166 at Hamble. The circular white panel behind the airscrew is a highlight of Westland constructed Type 166 seaplanes. The machine's serial, 9751, is stenciled to the trolley. The lower longeron footsteps of Westland built machines are well shown.

Above: No. 9752 at the Westland works. Westland placed the serial under the tailplane making it difficult to identify individual machines.

Right: Close-up of the wing folding arrangement and 200-hp Canton-Unné cowling on a Short 166 Type.

Left: A Westland 166 Type on the water. The vertical object in the observer's cockpit has not been identified. Also, note the white floats.

Right: The undercarriage has collapsed on this Westland Short 166 being recovered by HMS *Vindex*.

Left: No. 9760. a Westland constructed 166 Type, distinguished by the white circular cowl behind the airscrew. Note the curved separation-bars on the floats indicating it was built to carry a torpedo although no evidence of such ever taking place has been uncovered to date. Crewed by Fletcher and CPO A Blackwell, this Short has just returned from a North Sea patrol from HMS *Vindex*.

Right & Below: Bomb container built on HMS *Vindex* to streamline the bomb rack.

Below: Westland built 166 Type in late camouflage. This photograph was taken by B.L. Smith, USMC, after the US's entry into the war.

Specifications Short 166 Type			
Source	**1**	**2**	**3**
Span upper	54 ft	57 ft 3 in	57 ft 3 in
Span lower	40 ft	42 ft	42 ft 0 in
Length	39 ft	40 ft 7 in	40 ft 7 in
Height	11 ft 6 in	14 ft 0¾ in	14 ft 0 in
Dihedral	Nil	Nil	None
Chord		6 ft	6 ft 0 in
Gap		6 ft 3 in	6 ft 3 in
Stagger			None
Incidence			5°
Wings	544 ft^2	573 ft^2	
Ailerons (total)			64 ft^2
Tail plane			48.5 ft^2
Elevators			30 ft^2
Fin			15 ft^2
Rudder			19 ft^2
Loaded Wt.		4,580 lbs	4,580 lbs
Empty Wt.	2,700 lbs?		
Main floats Wt. (each)	250 lbs		
Fuel	105 gal.		
Oil	8 gal.		
Speed (max)		65 mph approx	
Engine	Canton Unné No.221		

Source:
1. Short Tractor 200 H.P. Serial No. 161 Log Book. TNA AIR1/2056/204/412/161.
2. J.M. Bruce data.
3. *Westland Aircraft 1915–1926 (The Westland Aircraft Collection Vol.1)*. Realvision Imagining Solutions Ltd. DVD.

British Aviation – The Pioneer Years, Putnam, UK, 1967. P.556.

4. *Flying Report for January 1916*. HMS *Ark Royal*, Salonika, 05 Feb 1916. TNA AIR1/649.
5. Cowlan, P.J. "The Short Admiralty Type 166 Seaplane – A Re-appraisal." *Cross & Cockade International*, Vol. 47 No. 1, 2016. P.10.
6. War Experiences of W/Cdr Bromet 1914–1918. (2nd Course) TNA AIR1/2387/228/11/30.
7. HMS *Roberts* and *Raglan* were monitors. These ships mounted 14-in guns and they *were one of the first types of warships to be built specifically with the operation of these aircraft as a prerequisite*. Cowlan, P. Op Cit. P.14.
8. Cowlan, P. *Op Cit*. P.16.
9. *Flying Report for January 1916*. HMS *Ark Royal*, Salonika, 05 Feb 1916. TNA AIR1/649.
10. Bruce, J.M. *British Aeroplanes 1914–1918*, Putnam, UK, 1957. P.482. All the Shorts-built 166 Type served on *Ark Royal*.
11. 'Flying Report for January 1916,' and 'Note on The Years Work of H.M.S. "Ark Royal" 1915–1916. HMS *Ark Royal*. 5 Feb 1916. TNA AIR1/649/17/122/48.
12. *Ibid*. HMS *Havelock* was an Abercrombie-class monitor. She sailed for the Dardanelles in June 1915, and returned to the UK in January 1916.
13. "Flying Report for January 1916." *Ark Royal*, 05.02.1918. TNA AIR1/649. The Greek Prime Minister had evacuated the Aegean islands of Tenedos and Lemnos that Greece had occupied after the First Balkan War. Turkey had not given up its claim to these islands and by vacating them and letting the Allies occupy them to set up naval and air bases the Greeks had control without troop commitments. Greece retained control of Tenedos until 1923 when it reverted to the new Turkish Republic.
14. *Aeroplane Log Book, Admiralty No.161*. TNA AIR1/2056/204/412/161.
15. Sturtivant & Page indicate that No. 161 was to have a land chassis installed around September 1916. No

Short No. 161 Record of Flights

Date	Duration		Name & Rank of (a) Pilot (b) Passenger Oil in tank	Oil in Tank	Petrol in Tank	Remarks
23.04	-	20	(a) Capt Kilner (b) Lt Park	6	70	Machine very tail heavy.
"	-	32	"	6	70	Machine climbed well, 2000 ft in 10 minutes. Adjustments made to tail correct.
"	-	32	"	6	70	W.T. test. Machine flew well. WT not working, leads adrift.
25.04	-	10	"	8	75	Started on reconnaissance, relived Spindle in oil pump sheared.
"	-	58	"	8	75	Reconnaissance of Gaba Tepe. Machine flew well. Baragraph broke at 6500 ft. Climbed 3000 ft in 17 minutes.
26.04	-	34	"	8	75	Reconnaissance Gaba Tepe district
"	1	03	"	8	75	General Reconnaissance of Gaba Tepe with W.T.
27.04	-	52	(a) Capt Kilner (b) Lt Cmdr Hornby	8	75	Reconnaissance of Gaba Tepe. Engine cut out at 3000 ft.
"	2	16	(a) Capt Kilner (b) Lt Brown	8	75	Reconnaissance of Gaba Tepe, with main objective of spotting batteries controlling fire onto them. Nothing seen. Height 5400 ft.
30.04	1	20	a) Capt Kilner (b) Lt Davies, RFA	8	75	Reconnaissance of district of Gaba Tepe. Height 6200 ft.
"	2	11	(a) Capt Kilner (b) Lt Hornsby	8	75	Spotting for HMS *Triumph* & general reconnaissance. Height 6000 Ft.
01.05	1	10	"	8	75	Spotting for HMS *Prince of Wales*.
"	1	20	(a) Capt Kilner (b) Lt Park	8	75	Spotting for HMS *Queen Elizabeth*.
"	2	45	"	8	75	Observe if possible batteries firing, & also to keep down gun fire during infantry attack. Height 6400 ft.
02.05	1	10	"	8	75	General reconnaissance of Gaba Tepe. Height 5000 ft.
"	-	57	(a) Capt Kilner (b) Lt Davies, RFA	8	75	Observe batteries & if possible keep down gun fire during in-fantry attack. Height 4500 ft, bad climbing weather.
03.05	-	57	(a) Capt Kilner (b) Lt Park	8	75	Spotting for HMS *Queen Elizabeth* Height 3500 ft.
04.05	-	49	(a) Capt Kilner (b) Lt Davies, RFA	8	75	Reconnaissance of Gaba Tepe & district. Height 3400 ft. Machine hit in carburettor by rifle fire.
05.05	2	26	(a) Capt Kilner (b) Lt Park	8	75	Proceeded to spot for HMS *Canopus* on battery in 177 M got up Dardenelles (sic), but recalled to spot for HMS *Triumph* on Maidos. Carried out this spotting. Maidos on fire. Height 6400 ft.

Short No. 161 Record of Flights						
Date	Duration		Name & Rank of (a) Pilot (b) Passenger Oil in tank	Oil in Tank	Petrol in Tank	Remarks
07.05	1	10	"	5	80	Spotted HMS *Prince of Wales* on to Maidos. Height 4000 ft. Fourth & successive shots all hit. Town on fire.
11.05	2	00	(a) Capt Kilner (b) Lt Strain	7	75	Reconnaissance from Chanak to Gallipoli. Height 5400 ft.
13.05	-	13	(a) Capt Kilner (b) Lt Park	7	75	In getting off in a long swell, after chassis struts carried away.
13.05	1	05	(a) Capt Kilner (b) Lt Strain	7	75	Reconnaissance of Gaba Tepe. Very bad for climbing. Height 3200 ft.
20.05	1	00	(a) Capt Kilner (b) PO Marchant	7	75	Light bad for observation. Returned owing to darkness & engine missing.
25.05	2	35	(a) Capt Kilner (b) Lt Cmdr Brodie	7	75	Reconnoitring for submarine. HMS *Triumph* sunk.
29.05	2	19	"	7½	90	Submarine reconnaissance area. Engine running well.
03.06	-	56	(a) Capt Kilner (b) Lt Park	7½	90	Observed fire of HMS *Talbot* on Banyak Fort & Village. Very bad climbing day owing to bumps. Engine running hot, losing water.
25.06	-	15	(a) Capt Kilner (b) Lt Strain	7	90	Flew from HMA *Ark Royal* to Base.
26.06	-	25	(a) Capt Kilner (b) Lt Park	8	100	Machine out of truth, would not get off water. Right wing always touching water.
27.06	-	10	a) Capt Kilner (b) CPO Goodman	7	90	Test flight, machine in truth, but engine overheated badly.
"	-	10	(a) Capt Kilner (b) LMS Wood	6¾	87	Test flight engine still overheating.
29.06	-	15	a) Capt Kilner (b) Lt Allsop	6	80	Test flight, engine gave 1170 revs & still somewhat hot, but fit for service.
05.07	-	15	(a) Capt Kilner (b) CPO Jones	5½	58	Test flight, all correct.
05.07	-	20	(a) Capt Kilner (b) Lt Park	5½	58	Spotting trip, had to come down elevator jambed at 80 ft. Floats damaged.
08.07	2	15	a) Capt Kilner (b) Lt Strain	5	60	Spotting for HMS *Humber*, very bad atmosphere, machine never xxxxx 4000 ft in 21 min. 1275 revs engine going very well, note pump packing conked out.
11.07	-	40	"	5¾	60	Searching for submarine. Nothing seen. Engine running well.
15.07	-	45	(a) Capt Kilner (b) Lt Park	5	60	Reconnaissance & taking photographs Engine running on full throttle. Taxied for 35 min. Engine time 1.20.
19.07	-	23	"	5	60	Observation & photographic flight. When HMS *Roberts* was to fire at Chanak. Radiator leaked badly. Returned. Time in air 15 min Height 1200 ft.

Short No. 161 Record of Flights						
Date	**Duration**		**Name & Rank of (a) Pilot (b) Passenger Oil in tank**	**Oil in Tank**	**Petrol in Tank**	**Remarks**
21.07	1	03	"	5	60	Proceeded to spot for HMA *Humber* on to Niados. W.T. jambed. Height 4600 Engine time 1.10.
22.07	1	10	(a) Capt Kilner (b) Lt Park	5	60	Reconnaissance of 249 M, also taking photos. Height 5400. Very bad flying day, bumps everywhere.
23.07	1	50	(a) Capt Kilner (b) Lt Strain	5	60	Spotting for HMS *Abcrombie* on Chanak. Quite satisfactory. Height 5500 Engine ran well. Engine time 2m 3s. Landed alongside HMS *Ark Royal* & hoisted in.
24.07	-	15	(a) Capt Kilner (b) Cmdr Woolbridge	2	20	Returned to Base. At 400 ft engine cut out, water in petrol, but picked up again. Revs 1250.
29.07	1	40	(a) Capt Kilner (b) Lt Strain	5	60	Taking photographs of square W4. Afterwards spotting for Monitor *No.34*. On to Maidos. Engine ran well, but slight vibration once for period of 5 min. Height 3500–4000.
30.07	1	17	"	5	60	Taking photographs & reconnaissance of 104 Bi. Engine occasionally vibrated. Revs 1275. Height 4200.
10.08	-	15	(a) Capt Kilner (b) AH Summers	5	60	From Base to ship to have new port float fitted. Tried Integral propeller. Engine gave 1350 revs.
10.08	-	15	(a) Capt Kilner (b) Lt Park	4¾	58	From ship to base.
12.08	-	15	(a) Capt Kilner (b) Lt Allsop	4	55	From Base to ship to have new planes fitted. Flying all correct.
14.08	-	15	"	3¾	52	From Ship to Base. Flying all correct.
15.08	2	00	(a) Flt Lt Douglas (b) Flt Lt Pulford	5	60	Spotting for HMS *Edgerton* S&M batteries. Some photos taken. Max height 4500 ft.
16.08	1	50	(a) Capt Kilner (b) Lt Park	5	60	Spotting for HMS *Theses* onto 5" batteries. Height 5000 ft. Engine ran well. Bottom of both floats tore off on landing. Machine salved.
01.09	-	20	a) Capt Kilner (b) AM Wren	3	30	Base to *Ark Royal*. Engine ran well. 1230 revs.
05.09	-	25	(a) Capt Kilner (b) Lt Pulford	5	70	Proceeded to spot for HMS *Thesus* returned owing to vibration & engine knocking, due badly balanced propeller, causing engine to overheat.
09.09	-	30	(a) Capt Kilner (b) Cmdr Robinson	5	60	Reconnaissance of left flank batteries at Helles. Returned engine missing, due Worse lubrication. Climbed 1000 ft in 2 min.

Short No. 161 Record of Flights						
Date	**Duration**		**Name & Rank of (a) Pilot (b) Passenger Oil in tank**	**Oil in Tank**	**Petrol in Tank**	**Remarks**
16.09	-	15	(a) Flt Lt Douglas	5	55	Flight from ship to base. Engine giving 1250. Ht 1000 ft.
16.09	-	25	(a) Flt Lt Douglas (b) Flt Lt Dunning	5	60	To spot for *M.32*. Wireless found faulty on test, causing return. Adjusted successfully just before arriving back but engine missed badly owing to faulty distributor busk
18.09	1	00	(a) Flt Lt Douglas (b) Lt Park	5	60	To spot for HMS *Theseus* onto D5 battery. Good shooting. Could not climb above 2200 ft. Cause unknown. Revs up to 1300.
18.09	-	10	(a) Flt Lt Douglas (b) F Cmdr Bromet	4	50	Taxy in dark (but moonlight) to test lighting apparatus.
19.09	-	20	a) Flt Lt Douglas (b) Lt Park	4	50	Base to ship.
27.09	1	40	(a) Capt Kilner (b) Lt Dunning	5	60	Spotting for 1732, owing to sea she was unable to fire. Machine climbed badly. Reached 4100 in 1 hr.
25.10	-	55	(a) S/Lt Morrison (b) Lt Rylands	5	60	Practice flight.
05.11	2	45	(a) Sub Lt Morrison (b) Lt Brown, RNVR	7	70	Reconnaissance of the Gulf of Smyrna. Took photographs of town of Smyrna. Maximum ht only 2000 ft.
06.11.	1	30	(a) Sub Lt Morrison (b) Lt Parke, RNVR	7	70	Fifteen photographs taken of the Bay of Aivali. Returned early owing to oil pump leaking & covering machine with oil.
10.11	2	20	(a) Sub Lt Morrison (b) PO Marchant	7	70	Patrol of the Gulf of Salonica from the harbour of Salonica as far south as Kisslol Pt for submarines. Observation very difficult owing to thick clouds at 500 ft.
14.11	1	05	a) Sub Lt Morrison (b) Lt Strain, RNVR	7	70	Reconnaissance & photography from Tugla Pt to Salonika along coast.
15.11	-	50	"	7	70	Reconnaissance & photography of C. Kara, etc. Returned owing to bad weather.
19.11	-	45	"	7	70	General reconnaissance of Salonika.
21.11	-	55	(a) Sub Lt Morrison (b) Lt Brown, RNR	7	70	General of Salonika.
22.12	1	00	(a) Sub Lt Morrison (b) Lt Pallissen	7	70	Examination of Salonika defences by Lt Pallissen, gunnery Lieutenant, HMS *Exmouth*.
22.12	-	01	(a) Sub Lt Morrison (b) Flt SubLt Whetnull			Trial flight for Sub-Lt Whetnull. Engine cut out when machine was 10 ft off water.
22.12	-	10	"	7	70	Trial flight for S-Lt Whetnull, returned owing to badly missing engine.

Short No. 161 Record of Flights						
Date	**Duration**		**Name & Rank of (a) Pilot (b) Passenger Oil in tank**	**Oil in Tank**	**Petrol in Tank**	**Remarks**
23.12	-	45	(a) Flt SLt Whetnull (b) Eng Cmdr Woolridge	7	70	Practice flight.
28.12	-	28	(a) Flt SLt Whetnull (b) Lt Strain	7	70	Reconnaissance of Kara & Tuzla Pt. Three photographs taken of D fort.
28.12	-	35	(a) Flt SLt Whetnull (b) Sub Lt Enington	7	70	General reconnaissance.
1916						
19.01	-	01	(a) Flt Lt R Whithead (b) Sub Lt Erington, RNR	7	70	Attempted trial flight engine cutting out.
20.01	-	23	a) Flt Lt R Whithead (b) CPO XXXX	7	70	Trial flight after had undergone repairs.
25.01	-	45	(a) FLt R Whitehead (b) Sub Lt Erington, RNR	7	70	Reconnaissance flight round forts on Kara & Tugla points, three photographs taken.
08.02	1	00	(a) Flt Lt R Whitehead (b) CPO Munson	7	70	Wireless flight to tune new shore sets onto D&P wave.
03.03	-	26	(a) Lt (b) PO Munson	7	70	Flight to test telepand receiver in accordance with V.A.'s orders.
07.03	-	15	a) Lt (b) Ass Paymaster Diver	7	70	Flight in order cinematograph pictures could be taken in accordance with R.A. Exmouth's orders.
21.03	-	25	(a) SLt Whetnall (b) Mr Ellison	7	70	Test flight.
11.04	-	36	(a) SLt Whetnall (b) Lt Tonington, RNVR	7	70	Practice flight over Batchelor mirror. After third shot engine cut out partially had to return. Height 3,000 ft.
11.04	-	22	"	7	70	Second practice flight over mirror. Owing to bad conditions no lights were fired.
12.04	1	15	"	7	70	Practice flight over Batchelor mirror. Commenced at 2,000 ft but after 4 shots conditions becoming rapidly worse tried 1000 without improvement. Then climbed to 2,500 ft where air was settled. Practice fairly (oo) Eight shots were
29.04	1	30	(a) Flt Douglas (b) Lt Lord Torrington, RNVR	6	40	Search for submarine between Mudros & Castro. Two bombs taken. None sighted.
12.06	-	-	(a) Flt SLt Whetnall (b) Flt SLt Cross	7	40	Test flight also practice flight with Sub Lt Cross. (Machine is very slow in climb & heavy on controls.) (Note: No times recorded)

Short No. 161 Record of Flights						
Date	Duration		Name & Rank of (a) Pilot (b) Passenger Oil in tank	Oil in Tank	Petrol in Tank	Remarks
12.06	-	20	(a) Flt SLt Cross (b) AM Sommer	6	36	Practice flight.
21.06.1916 Transferred to No.2 Wing Mudros.						
Notes: Torrington became a PoW with Flt Lt E.J. Cooper when their Henry Farman F.27 3905 failed to return after a raid on Geraviz Seaplane Base on 01.12.1916.						

further trace of the machine has been discovered to date.

16. "War Experiences Sqn Ldr C.H.W. Pulford 1914-1918". RAF Staff College. TNA AIR1/2386/228/11/14. Note that the initials are in an incorrect order.
17. Sturtivant & Page, *Op Cit*. P.56. Pulford had crashed and wrecked Sopwith "Spinning Jenny" No. 1064 at Hendon on 25 May 1915.
18. The Army Zeppelin *LZ85* raided Salonika on the night of 31 January/ 1 February 1916, and again on the night of 4/5 March when it was brought down.
19. The *Ratesky* was a semi-dreadnought battleship built for the Austro-Hungarian Navy and based in Pola. It was too slow to operate with the *Goeben* and would have been unlikely to have escaped the Adriatic blockade.
20. "War Experiences Sqn Ldr C.H.W. Pulford 1914–1918". RAF Staff College. TNA AIR1/2386/228/11/14.
21. Diary H.R. Hadfield. FAA Museum, Yeovilton, D.Y.12. Box 3.
22. Sturtivant & Page, *Op Cit*. P.80 & 75.
23. Jones, H.A. *The War in the Air,* Vol. 5, P.374.
24. Cowlan, P. *Op Cit*. P.22.
25. Cowlan, P. *Op Cit*. P.22.
26. This section has been written in the main from Cowlan, P. *Op Cit*. P.23.
27. Minute of 03.04.1916. Under this was the note: "No objections so far as 'M' is concerned. 225 Shorts are being sent to EMed. To meet torpedo requirements." Signed T.C. Williamson. 04.04.1916. TNA AIR1/149/15/104.
28. The Y Section of the Admiralty Air Department only recommended acceptance of the *tender for supply and deliver 10 200 H.P. Canton Unne Seaplanes and spares* in the W/E 14 July 1917. TNA AIR1/337/15/226/247.
29. Penrose states that Arthur Davenport was tasked with the job of drawing up the *still incompletely designed bomber version* of the 166 Type. This must be a reference to the change from carrying a torpedo to carrying bombs. Penrose, H. *British Aviation – The Great War and Armistice*, Putnam, UK, 1967, P. 54.
30. Report of I Section, Admiralty Air Department. Date not given but filed with W/E 28.10.1916 material. The report concluded with: *No. 9756 (from the "Vindex" for repairs) is now ready for delivery*. TNA AIR1/337/15/226/247.
31. See Burns, I.M. & Nailer, R. 'HMS *Vindex* – Mixed Carrier 1915-1919,' *Cross & Cockade International*, Vol.17, No. 3 Autumn 1986. P.97, for a history of the *Vindex*.
32. TNA AIR1/2386/228/11/5.
33. Mottram, G. 'Norman Woodhead – Naval Aviator.' *Cross & Cockade Great Britain*. Vol.9 No. 2, 1978. P.80.
34. Minutes of the 13th Air Department Weekly Progress Meeting held on Tuesday 7th August 1917, at 11.30 a.m. TNA AIR1/270/15/226/113.

Chapter 8: The Short 827 Type

Right: No. 823 in a spot of trouble. The red/white rings on the upper surface of the top wing can be seen inboard of the ailerons. This is possibly the accident of 11 June 1915, when the starboard float was broken while taking off.

Under the incomprehensible Admiralty's designation scheme, the next products from Shorts were known as the 827 Type and 830 Type. These two seaplanes used the same airframe, but the 827 Type was powered by the 150-hp Sunbeam water-cooled V-eight engine, whilst the 830 Type was powered by the 135-hp Salmson Canton-Unné water-cooled radial engine. Three 830 Type machines (Nos. 828 to 830 – c/n S.116 to 118) were ordered under Contract No. C.P.57105/14, together with six 827 Type seaplanes with serial Nos. 822 to 827 (c/n S.122 to S.127). The 827 Type seaplanes were to have the 110-hp Sunbeam but were delivered with the 150-hp Sunbeam Crusader engine.

Under Contract No. C.P.57105/14/X.18826, three 830 Type machines were ordered with the serial Nos. 819 to 821 (c/n S.119 to S.121). No. 819 was delivered in December 1914, while No. 828 was delivered in November 1914.[1] Why the serial numbers were broken and why the machines were delivered out of sequence is not known. However, this is probably the reason that it is recorded that six of each type were originally ordered.[2]

The two types bore a close resemblance in structure and appearance to the Short 166. The machines were two-bay biplanes of conventional Shorts' construction with ash longerons and spruce struts with wire cross bracing. The upper wing extensions were braced by cables from kingposts above the outer set of interplane struts. Inversely tapered ailerons were fitted to the upper wing only. The struts were of steel tube. The mainplanes could be folded in typical Shorts fashion. A large, bulky radiator was mounted on top of the fuselage in front of the wings. The observer could be provided with a Lewis gun and small bombs could be carried below the fuselage. The fabric covering was linen and was given six coats of dope and two coats of varnish.[3]

The 827 Type began to appear in 1915. A number of Short 830 seaplanes were built, but on standardisation of the Sunbeam engine, the majority were Short 827 machines. The demand for the 827 Type was such that it was contracted out to the Brush Electrical Engineering Company, the Fairey Aviation Company, Parnall & Sons, and the Sunbeam Motor Car Company.[4] The batches were small. After Parnall became contractors for the Admiralty, they built two Short types, the first being the 827 Type, that was built in two batches, the first of twelve and the second of eight seaplanes. (See table of serial allocations at end of this chapter).

There was also a three-bay version of the Short 827 in which the lower wing was extended to be almost the same span as the upper wing. Nos. 8231, 8233, 8237 and 8561 are known to have had the three-bay arrangement.

When with No. 8 Squadron RNAS at Zanzibar, No. 3097 had the following modifications and additions to fit it for service: - Brackets for fitting W/T; alteration to the position of the compass; and fitting Verys Light rack. Other 827 Types would have had similar alterations once in service.

No. 3100 recorded that a Lewis gun mount was *fitted abaft the passenger seat.* Two 65-lb *bomb gears were fitted to the fore and aft line of the machine.* Bomb release levers, lever sights and a sliding bomb panel was fitted in the pilot's cockpit. No. 3097 recorded the following instruments: -

General Description of 8 Cylinder & 12 Cylinder Sunbeam Aero Engines

The January 1914 Sunbeam Aero Engine catalogue covered the 8 cylinder 150-hp and the 12 cylinder 225-hp engines.

Both models are constructed on similar lines throughout being of the V type. The Cylinders are disposed in "en bloc" castings in sets of four for the 8 Cylinder and sets of 3 for the 12 Cylinder.

The crankshafts are supported by bearings between each throw of large diameter. Each throw carries two connecting rods arranged side by side, the Cylinders being staggered to allow this. The shaft is hollow for lightness and the oiling system. The valves are of large diameter and are on one side of each set of Cylinders. A single camshaft actuates both sets of valves, this camshaft is situated between the Cylinders and is driven by spur gears.

Provision is made in front of the engine for the water pump and the magneto both of which are driven by spur gears. Lubrication is forced by a gear driven pump in the sump (which obtains its supply direct from an oil tank separated from the engine) to main bearings from thence it is forced through the hollow crankshaft to the big end bearings after which it is splashed. In the oil circuit a lead from the main oilway is taken to the tappets and rockers before emptying itself over the timing wheels and eventually into the base chamber.

Provision is made for fitting standard propellors in the nose piece f the engine, this is designed to contain a reduction gear of 2.1 the shaft running in two very heavy ball bearings. A special double thrust bearing is also introduced which enables the engine to be used for either tractor or propelled types of aeroplanes without any alteration to the design. Materials only of the very highest grade and selection are used throughout and the engines are built in a special department by the highest skilled workmen.

Therefore a very high grade of accuracy in the selection of material finish and workmanship is guaranteed.

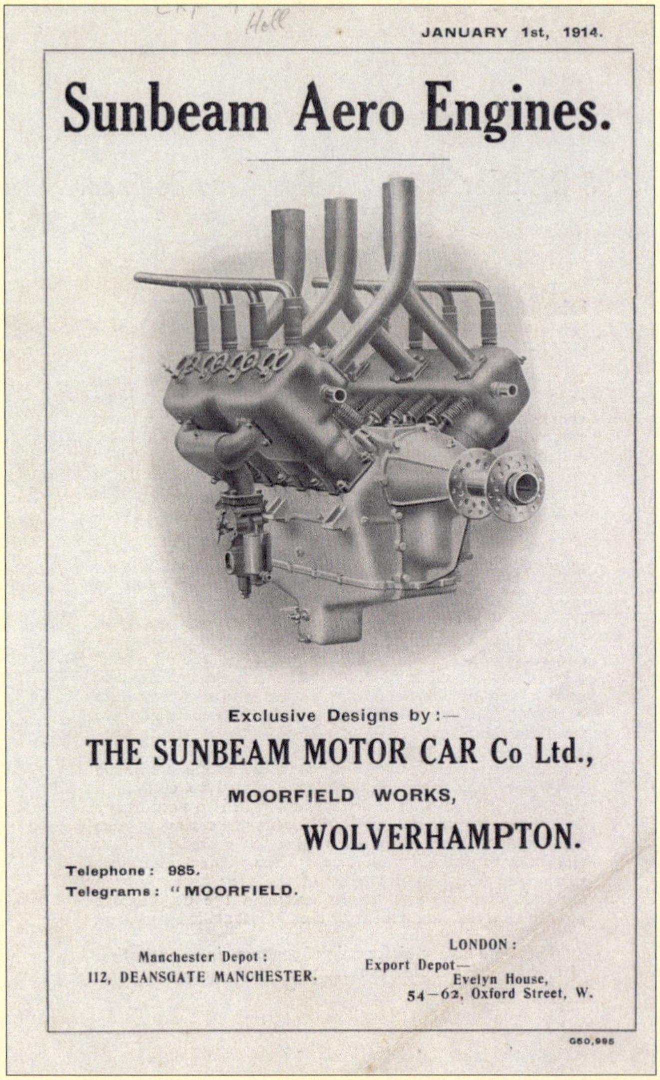

JANUARY 1st, 1914.

Sunbeam Aero Engines.

Exclusive Designs by:—

THE SUNBEAM MOTOR CAR Co Ltd.,

MOORFIELD WORKS,

WOLVERHAMPTON.

Telephone: 985.
Telegrams: "MOORFIELD.

Manchester Depot:
112, DEANSGATE MANCHESTER.

LONDON:
Export Depot—
Evelyn House,
54—62, Oxford Street, W.

G80,995

Above: The Sunbeam 150-hp engine is presented on the cover of the 1914 Sunbeam catalogue.

Sperry Compass, Short & Mason Altimeter, Elliott ASI, cross level, rev indicator, Shickarp inclinometer and Smiths clock.[5]

Home Waters

The Short 827 Type was used on Home Stations and overseas. They were based at Great Yarmouth, Calshot, Dundee, Dover, Killingholme, and the Isle of Grain for anti-submarine and mine patrols as well as training.

The renewed German U-boat offensive of 1916 saw a new seaplane station established at Portland late that year. Claude Hilton Keith was stationed on the Isle of Grain when one morning he and three others were ordered to fly over to Portland where the 3rd Battle Squadron was *more or less hemmed in by fear of hostile submarines*. Three 150 Sunbeam Shorts were accompanied by a 225 Sunbeam Short.

I was the first Grain machine to land at Portland Harbour where I found a 150 Sunbeam Short, Pilot Flight Lieut. Cooper, already moored astern of H.M.S. Dreadnought, which was then the Flagship of Admiral de Robeck, commanding the 3rd Battle squadron.

Cooper was given command of the five machines. Keith was attached to HMS *Dominion*. *As far as I was aware, we did no actual patrols during the seven days we were there, attached to the 3rd Battle Squadron. We made some flights, but experienced great difficulty in running the machines, owing to the absence of all proper facilities.*

Right: HMS *Campania* after her second conversion.

Left: Delivered to HMS *Campania* in April 1915, No. 823 displays the Union flag on the fuselage. It served most of the time with the *Campania*, being written off on 7 July 1917.

Right: Hoisting No. 824 on HMS *Campania*, 17 April 1915. The dark rear fuselage is probably caused by a shadow.

Left: HMS *Engadine* with a Short 184 Type floatplane with wings folded at the rear.

Below: No. 3065, on HMS *Vindex*, shows the bulky radiator that was a feature of all the Short in-line engined seaplanes. A presentation aircraft: "Britons Overseas No. 3," it was delivered to Grain on 16 September 1915. With HMS *Vindex* by October 1915. To the Nore War Flight in April 1916. It survived until June 1918.

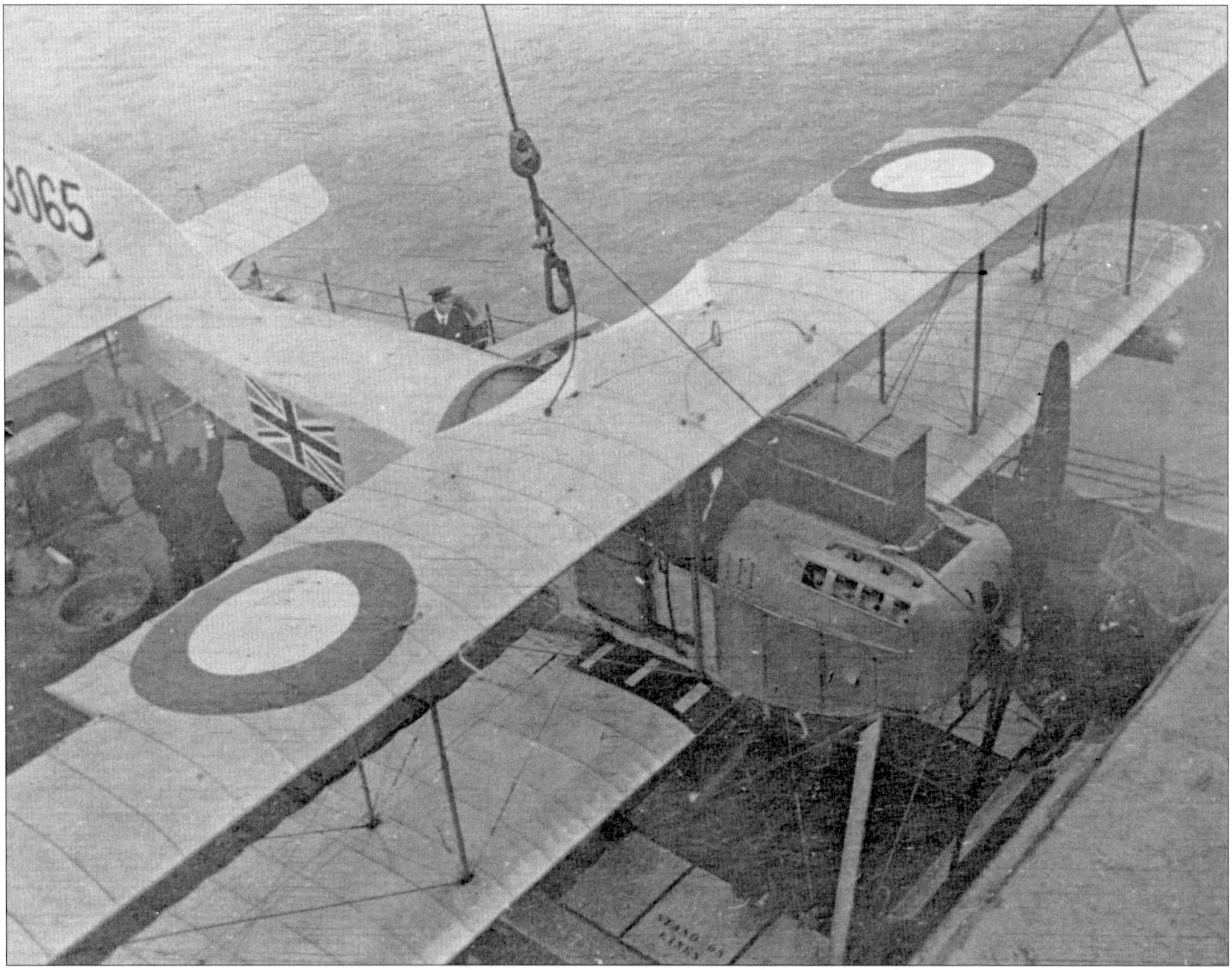

At the end of the seven days the Battle Fleet set sail from Portland for good. The Short floatplanes were ordered to be tied to the target slip in the dockyard. This they did the following morning.

The wind was so strong that it was necessary to have splints round the single acting ailerons in order to prevent them from damaging themselves, there being o hanger or shelter of any kind whatever. We remained three days and at the end of that time, three of us started back for Grain. One machine broke down at Dover, another at Westgate and, I believe I am right in saying, that only one got back to the Isle of Grain of the four original machines which flew to Portland. I quote this as an instance of the unreliability of the engine.

About three weeks later Keith was despatched from Grain to Calshot where the Patrol Flight was under Flt Lt C.L. Scott. There were three other pilots, Flt Sub-Lt Fox and Flt Sub-Lt Ross, and Keith. They had the *much begrudged use* of a hangar and of a dockyard shed.

The Naval authorities regarded us as unmitigated nuisances. We had no wireless for at least the first six months, I think I am right in saying. On the 150 Shorts we carried four 16 lb bombs and on the 225's we usually carried two 65 lb. bombs, or two 100's or one 100 and one 65. We had no set patrols, but in the absence of hostile reports, we carried out large triangular patrols, wherever our fancy led us. Two patrols a day was about as much as we usually did in decent weather.

We had numerous forced landings, and lost a considerable number of machines. Flight Sub-Lieut. Fox[6]*, who was, unfortunately, lost at sea just after Flight Lieut. Waugh replaced Flight Lieut. Scott as O.C., station, had some nineteen forced landings, I believe. He was, in my opinion, the most resourceful and full out seaplane Patrol Pilot that I have ever met or heard of. On one occasion on a forced landing, he constructed a drogue from the cushion cover of the machine, which acted very successfully. He did some very long flights, the best of which was, I think, about 5½ hours on a 225 Sunbeam Short.*

One of the most striking incidents of the patrols was the occasion on which Flight Sub-Lieut. Ross, after bombing and missing a hostile submarine, crashed and was being salved by a "Q" ship, when the submarine attacked the "Q" ship which had to cut the seaplane adrift and Lieut. Ross, was a close eye-witness of the successful retaliation by the "Q" in which the Hun was sunk and some twelve prisoners, including the Commanding Officer were brought in to Portland.

Flight Sub-Lieut. Ross was the first Portland pilot to bomb a submarine, but, unfortunately, he did not hit it.[7]

Ross's adventures with submarines did not end there. Flt Sub-Lt John Fredrick Ross, with AM J Redman, DSM, (later Sgt Major Redman), left Portland Harbour on 30 November 1916, in Short 184 No. 8379 at 1205. At 1250 they sighted a Norwegian steamer heading about West and at *1225 I sighted "oil track" on the water about a mile to the northward of the Norwegian steamer. The "oil track" led to the West and at 1305 the end of the "oil track" was sighted. Oil was still coming to the surface.*

From 1305 to 1325 we kept the end of the "oil track" in view by circling above it, then we tried to warn the Norwegian, sending the following by lamp: "Submarine one mile North of you"; she did not understand evidently for she circled around and then proceeded on her course. I again picked up the end of the "oil track" and kept it in sight. At 1340 the tank steamer before sighted was again seen coming down from the North, at 1350 we started towards her to warn her, but when we had gone about three miles we saw the submarine about three miles astern of us and about a mile away from the Norwegian.

We immediately turned and headed for the submarine, at 1353 we were over the submarine and dropped a 65-lb. bomb.

The submarine was almost completely submerged when they dropped their bomb which missed the target. They waited keeping the oil stains in sight until 1415, when the tank steamer was now much closer, about two to three miles, so Ross decided to alight.

We signalled her but could not get her reply as the yellow and red flags did not show up very well against her upper works, so I landed and spoke to the Commander who said that if I would direct him he would drop a depth charge over the submarine.

I got off fairly well, but when about 150 feet in the air the machine stalled and side slipped – just as she was beginning to right herself the wing tip touched the water and the whole machine slewed around carrying away the main, tail and wing floats. The machine was awash at once.

The tank steamer which turned out to be H.M.S. "Q.7." under Cdr F.H. Grenfell, *sent a boat to the rescue, and herself came near and passed a rope through the lifting sling, in order to hoist the wreck on board; this rope carried away, but grappling irons were made fast and she was about to be hoisted inboard when the submarine appeared and commenced to shell the ship. Seaplane No. 8379 was cut adrift and "Q.7." proceeded as if an ordinary merchant vessel.*[8]

According to R.M. Grant, (*U-Boats Destroyed*, Putnam, UK, 1964), the UB-19 fired a warning shot at the Q-Ship HMS *Penshurst*. The vessel followed the usual Q-Ship method and when the submarine was within 1,000 yards, the crew appeared to abandon ship. The German commander, Niemeyer, was proceeding towards the stern to read the ships name when he became suspicious due to its battleship grey paint scheme. Just as he noticed there was no name and he prepared to dive; the Q-Ship opened up at 250 yards. The UB-19 was subject to 83 rounds and sunk. Niemeyer was amongst the 16 survivors. Ross's observer, Redman, was reported to have *maintained extraordinary coolness throughout the proceedings.*[9]

Ross unsuccessfully bombed a submarine on 20 December 1916, in Short 827 No. 3327. Ross also lost Short 184 No. 8071 and Wight No. 9846 in separate incidents in his career but escaped each time.[10]

The German High Seas Fleet conducted what became known as the Lowestoft Raid on 25 April 1916. The German cruisers bombarded the coastal towns of Lowestoft and Great Yarmouth in order to lure the British ships out to where the remaining High Seas Fleet was waiting where they would be defeated in detail. As a result, Grain sent out three 827 Type seaplanes to attempt to overtake the German cruisers and bomb them. The Shorts were Nos. 3102 (Flt Lt Halifax), 3103 (Flt Sub-Lt Malet[?]) and 3106 (Flt Sub-Lt Cross). As no observers were recorded it is assumed they flew solo. The trio left Grain at 0730 and were in the air for 2½ hours, flying out into the North Sea and then back to Felixstowe. *No hostile ships observed.* The trio returned to Grain late that afternoon.[11]

Above: No. 3067 was another presentation aircraft, the 'Hong Kong Britons No. 3 – Britons Overseas No. 8.' It appears to be brand new in the photograph judging from the sheen on the fabric surfaces. Delivered on 17 September 1915, this machine was lost while on patrol on 14 December 1915; Flt Sub Lt A.J. Nightingale being made a PoW.

Even small errors could have drastic results as the fate of Sunbeam manufactured No. 8639 shows. On 12 October 1916, Flt Sub-Lt Mostyn-Lewis left Grain on patrol at 0630. with P.O. William Herbert Hodgson as observer. It seemed that the pilot stalled his machine after rising from the water this being followed by a side-slip and nose-dive from a height of about 75 feet, the machine crashing into the water of the River Medway opposite the Grain pier. The pilot managed to get clear of the wreckage and was taken ashore suffering from shock. Hodgson did not get clear and was under water for nearly two hours and was quite dead when taken out of machine.[12] No. 8639 had been delivered to Grain on 10 May 1916.

On 12 March 1917, Dundee Short 827 No. 8645 was sent out after a report of a submarine off Bell Rock in the afternoon. At 1540 Flt Sub-Lt McNicholl with LM[13] Jenkins took off and sighted a submarine about two miles north-east of Bell Rock. She submerged before the pilot could get into position to attack. McNicholl continued to patrol in the vicinity and about five o'clock sighted the submarine again. The observer released four 16-lb bombs that fell in a direct line with the submarine about 100 feet apart. The first two dropped short, but the last two appeared to fall very close to the submarine as it was submerging, however, it was not possible to see if it had been hit.[14]

German East Africa

In March 1916, three 150-hp Short seaplanes, Nos. 3096, 3097 and 3098, had been brought to Zanzibar on the *Clan MacPhearson.* Geoffrey Moore supervised the unloading of the packing cases containing the seaplanes and had to report to the Senior Naval Officer (SNO) who was the captain of the ill-fated *Pegasus,* Jack Ingles. Moore was given instructions to find a building with an outsized flagstaff alongside it. He found such a building and was led through a series of opulent rooms and eventually was asked to wait in a small room. He was duly surprised when the Sultan of Zanzibar entered. He had mistaken the palace for the SNO's offices![15] The unit these seaplanes were to serve with in the East Africa campaign was later named No. 8 (Naval) Squadron under Flt Lt J.T. Cull who we met in Chapter 5.

The Squadron's base was some six miles south of the town at Chukwani, where a disused palace was made available to the RNAS. Leading Seaman Edward R.W. Lincoln was in charge of stores and recalled that the *palace itself is an immense, 3-storied building which consists of a spacious veranda running round on all sides, is supported by huge stone pillars and these give a very imposing appearance to the palace. The inner portion of this first floor consists of 2 very large and lofty central halls, marble paved and adjoining these on either side are a number of large bedrooms. The majority of the fellows prefer to sleep out on the veranda, not unnaturally. The ground floor consists of 2 large central halls, again marble paved, with various offices and rooms leading off on either side. The rear halls together with the 2 rooms on either side serve as our mess quarters, etc. The top of the building is one immense attic which at present is put to no use whatsoever. Nowadays, the place looks neglected and dilapidated but in the days of its glory it must have been very fine.*[16]

The seaplane base was on a nice beach of hard sand about half a mile from the old palace.

Moore carried out the check flights after the machines were erected *on the quays in a swelter never to be forgotten.*[17] He was to fly the first machine erected to Chukwani. The whole population of Zanzibar turned out to see the machine take to the air. After testing the controls Moore took off and climbed steadily to 2,000 feet as he was taking no chances with the tropical climate. He pushed the control forward but the machine kept climbing despite all his physical exertions. He reached 5,000 feet, which he recorded as the greatest height he ever took one of the Shorts in the tropics. He was sure it would stall and as the Shorts were usually tail heavy, would get into a spin and that would be the end of them.

The Short stalled like a *perfect lady* and they volplaned down to alight on the water. The investigation found that the elevator had been put on upside down. The kingposts gave more leverage to pull the nose up, but none to put it down. When testing the controls, he had not known that the elevator fell under the influence of gravity and not under his control. On the second machine he had an engine fire in the air but this was put out by the onboard extinguisher.

On 26 March, Lincoln recorded that *One of our seaplanes flew from Zanzibar to our station here this morning, Lt Moore being the pilot. Unfortunately, later on, in the afternoon, Lt Moore had a mishap or rather a wing gave way and the machine dropped some 60 or so* (feet) *into the water. A good job it was not a land bus.*[18]

To assist the Cape Squadron's activities in German East Africa, the Sunbeam Short seaplanes would be embarked on to ships of the Squadron – the armed merchant cruiser HMS *Himalaya*, the balloon ship HMS *Manica* that could accommodate a seaplane aft, and the armed liner *Laconia.* The machines were lashed to the deck near the derrick used to lift them aboard and onto the water. The airscrew would be covered by a canvas protective cover. The ship would stay hull down from the land so that the Germans did not know where the aircraft were coming from. A cruise of seven to ten days off the coast would see the machines returned to Chukwani where a permanent hangar was erected for maintenance.

Lincoln recorded that *the actual erection used by us out here for our big seaplanes is termed a Bessoneau. These huge portable tents are tolerably easy to erect and take down. At a pinch* (it) *is capable of accommodating 5 Short seaplanes with their wings folded. The floor of the tent consists of a sandy surface on top of firmly pressed down soil and the machines are moved about on specially constructed trolleys. It is a very wearisome business working with seaplanes, especially if the tide is unfavourable and the machine has to be pushed a long way over soft sand. A depth of at least 2 ft 6 in of water has to be reached before the plane can be safely launched off the trolley. It is an even more tedious business getting a machine in when the tide is out., but the fellows all work with a will and more often than not we enlist the services of the native guard which dwells in the immediate neighbourhood of the hangar.*[19]

The Shorts would conduct reconnaissance, bombing flights and observation for naval bombardments, commencing on 10 April 1916, with Nos. 3096 and 3097 on the *'Laconia' an old Cunarder,* (that) *has been specially fitted out as a seaplane carrying ship and her armament consists of eight 6 inch guns.*[20]

R.D. Pershore kept a diary of these events:

Today I start out with an expedition for Tanga which we are going to bombard tomorrow. I take a supply of goods with me and we go aboard the Leconia (sic) *at 2 P.M. with 2 Seaplanes 6 officers & 14 men. Tonight as I write this we are sailing towards Tanga's German base and we have orders to be ready at 5 tomorrow morning.*

Tues. April 11 1916

7a.m. We are sailing into action now. 10.30 am. Our two seaplanes are launched one loaded with 16 lb bombs and the other with photo gear. 11 am Bombardment commences by the 3rd Class Cruiser 'Pioneer' and a gun-boat 'Thistle' the Cruiser 'Vengenance' (sic) *is also lying near but does not take part as there is no reply to the ships. The seaplanes are hit by rifle shot but return at 12.*[21]

The object was to get the enemy's guns to fire, then the *Vengeance* would have opened up with her 12-inch guns, however not a single shot was fired at any of the ships.[22]

Lincoln made many interesting observations in his diary. The following entry was made after the bombing raid on Tanga.

During wartime, most people associate aeroplanes and airships with bombs; and for very good reasons too. Undoubtedly, one of the chief functions of aeroplanes and seaplanes during hostilities is the dropping of these highly destructive missiles on fortified works, stations, warships and even among troops….the largest type of bomb we have with us is the 65lb HE bomb. These bombs are filled with amatol poured into the body of the bomb in a brown liquid state which by degrees solidifies...the 16lb HE bomb, of which we have a great number, looks exceedingly diminutive when placed alongside a 65lb bomb. Nevertheless, these little yellow-painted, pear-shaped, fellows are very deadly when they strike buildings or some other hard surface. The 16lb HE bomb is also filled with amatol and like the big brother has inside it a long exploder with one detonator at the tail end. Only direct action is available on these bombs…..the 15lb 'Carcass' incendiary bomb is another type used by this squadron. Similar in size and shape to the 16lb HE it is filled with a highly inflammable powder, the

In his diary Edward Lincoln recorded the opinions of his officers which are enlightening:

Generally speaking, the standard of efficiency among *officers of the RNAS is not particularly high, this is, of course, from the point of view of all round, thorough, capability. As pilots pure and simple little fault can be found with them, but it would seem that the flying man is almost a distinct type; he has all his wits about him while flying but loses them when he comes to earth. As an executive officer he is an utter failure. One explanation for this state of affairs is that pilots for the Services are drawn from the ranks of youngsters, many of them mere boys. Again, flying makes no special demand for men with brains in any way above the average. The qualifications for becoming an air pilot are much on a par with those required for speedy motoring or, more appropriately, speedy motor cycling – a cool head and a decent nervous system. Thus it is that airmen are drawn to a very considerable degree from that type of young gentlemen commonly known as 'Knuts', particularly that type which before the war spent most of its time flying about our main roads in high powered cars and motor cycles.*

Most fortunately, however, there is another better type than the foregoing which provides the Air Service with the very best officers and pilots and this Squadron is lucky enough to be officered by 2 or 3 men of this sort.

Sqn Cdr Nanson, a man of about 38 or 40, is our CO. In appearance, clean shaven, sharp featured, with a discerning eye and a thin-lipped, somewhat large mouth which is for ever shaping itself into a most disconcertingly sardonic grin. Very few people know what to make of Nanson's grin. If a man were to tell him he had deliberately set fire to a 'plane, Nanson would grin for about 10 seconds though there would be a deadly look mixed up in it. I have never seen or heard the fellow laugh; he never gets beyond his compressed-lip grin. As a good steady pilot, Commander Nanson ranks among the leading airmen of the RNAS. During his whole flying career which dates back to pre-war days, he has never smashed up a machine. Amongst the good works done by him during the war, mention must be made of his participation in the great Cuxhaven air raid when he made a successful journey there and back and dropped bombs with some effect. For some time he was CO at Dunkirk Air Station when he did many flights under heavy anti-aircraft fire. I first came into contact with the CO at Dundee Air Station where he was at the head of affairs. Nanson is a man who possesses a good deal of personality of a desirable kind and the men work very hard for him. He has the happy knack of evincing signs of pleasure when work has been well done, without the use of words. Unlike many COs in the Air Service, Nanson has a very keen sense of duty and takes a genuine interest in everything which appertains to the squadron. 'Duty before pleasure' is undoubtedly the CO's motto. He has gained his promotion from Flight to Squadron Commander since being out here and now wears three rings. It is my earnest hope that I may remain under the command of Nanson all the time I am in the Air Service. I have experienced other COs and like them not.

(Eric Roper Curzon Nanson. His RAC Certificate was granted on 15 October 1913. Lincoln is incorrect in that Nanson did not take part in the Cuxhaven Raid. Other parts of his history of Nanson are not quite correct, but Nanson made a good impression on his men.)

Flight Lieutenant (Edward Rowland) *Moon is our second in command and has most to do with the working of the station. Thirty years of age, he has had long and wide experience with most kinds of machines and at the present time he also holds place among the crack pilots of the RNAS. As an executive officer, he fulfils his duties admirably. There is nothing about Moon at which one can grumble and to me he has been exceedingly decent from the very commencement of the Expedition. Many a long yarn have I had down in the armoury with Moon, and much interesting information he has given me concerning his flying experiences. He is a man of unusual frankness and gentleness and one of the highest tributes I can pay him is to state that more often than not he forgets that he is an officer.*

(Sqn Ldr Moon, DSO and bar, was killed on 29 April 1919, taking off in the Felixstowe Fury triplane flying boat on a trip to Egypt. Born on 8 July 1886, he spent his early life at Southampton. In 1909 he carried his interest in aviation further by building a steel monoplane akin to the Santos Dumont Demoiselle, but the JAP motorcycle engine he used had insufficient power to get the craft airborne. He established a successful motor boat and car business as Moonbeam Ltd. In 1914 he joined the RNAS as a Flt Lt in August and was posted to Hendon under John Porte. He passed his tests for his certificate on 10 October 1914. Posted to East Africa in February 1916 with No. 8 Squadron, he was then involved in the incidents recorded here. He was awarded the DSO for his constant flights over the enemy coast and his great coolness and resource. Moon was described in his obituary as a modest man, clever, gallant and unassuming. He was attracted to Socialism, though in later life, he was less exuberant about it. He was an inspiring companion and a wise friend.)

Flight Lieutenant (William Geoffrey) *Moore (is) an altogether excellent young fellow of three and twenty and probably one of the finest pilots in or out of either of the Air Services. Moore, who is most extraordinarily like my friend Lt Fitzgerald, possesses a remarkably charming manner. Quiet of speech, entirely unassuming, he is without exaggeration one of the best fellows in the Service. Moore is an exceedingly keen airman and he is never satisfied unless he gets better results out of any machine than previous pilots who have handled them. Out here, he it was that made the test flights on all our seaplanes, on one of which he attained the record height of 6,500 ft. He has looped the loop dozens of times during his career and takes almost boyish delight in performing all kinds of aerial stunts. We are most fortunate in having such a valuable man in our squadron.*

Flight Lieutenant McLean (is) another young and first-rate pilot but entirely useless in any capacity except as a consumer of whisky. He has always been most decent to me but among the officers he is not at all popular, being somewhat foul-mouthed and a bit of a bounder. Nevertheless, I like old Mac and he is covering himself with glory out here having made a great number of excellent flights over German East Africa upon which he has repeatedly been congratulated by Admiral Charlton.

Sub-Lieutenant (Cecil Walter) Fitzherbert is not a pilot but an expert in aero engines in which capacity he displays great efficiency. Again a really decent fellow. (Fitzherbert was awarded the DSC in recognition of the valuable services performed by him as observer in the RNAS during operations in East Africa. His reports on the country, enemy positions, etc., and his photographs have been of the greatest assistance to the operations.)

Lincoln was to accompany the RNAS work on land. See 'Royal Naval Air Service Operations in German East Africa 1914-1918; Part 3,' Cross & Cockade International, Vol.38, No.1, 2007, P.20.

Source: 'Extracts from the Diary of Leading Mechanic Edward Robert Wingrove Lincoln (F7939)' via P Dye. Entry for 1 Apr 1916.

latter being fired by a Very cartridge which goes off on impact. This bomb takes its name from the composition with which it is filled. Of large petrol bombs, we have a limited number, their holding capacity being about 2 gallons of spirit, but they are scarcely used, the carcass bombs being much more effective and more accurate in flight.[23]

On 24 April, Lt Moore managed to get one of the seaplanes up to 6,500 feet, which is very excellent for a 'Short' seaplane I happened to be up the gun tower at the time and the machine appeared very small though the drone of the engine was clearly heard. Moore is an extraordinarily good pilot and should prove a formidable foe should any aerial fighting take place out here - which is extremely unlikely.[24]

Gallehawk noted that on *June 27th one machine with the necessary personnel was embarked in H.M.S. "Himalaya," which had been fitted with a temporary hanger aft. From this ship air observation was provided for shoots by H.M.S. "Challenger" and "Hyacinth" on Daar-es-Salaan* (sic)*, Tanga and Kilwa Kiswani, the seaplane spotting by W/T. Daar-es-Salaan* (sic) *was bombarded by the whole fleet on July 19th, two seaplanes and one kite balloon being used for spotting. Three ships in the harbour and the railway station were hit, and various buildings were set on fire and destroyed. This was the last serious piece of work carried out by the squadron, as fever and dysentery contracted on shore was spreading among the R.N.A.S., so that our numbers were always below strength.*[25]

Work continued with a reconnaissance over Dar es Salaam on 4 July 1916, carried out by a seaplane and kite balloon. The seaplane was fired upon by machine guns but no other sign of life was seen.[26] On the 21st, Cull and Gallehawk spotted for HMS *Vengeance* during the bombardment of Dar es Salaam. Forced to land with engine trouble they drifted to Makatumbe Island located near the entrance to the harbour. Once the engine was restarted they spotted for HMS *Mersey*, firing at the *Feldmarshall*, a German merchantman. They were relieved by HMS *Manica* with her kite balloon.

Flt Lt Edwin Rowland Moon and AM F Wilmshurst flew No. 3097 on a reconnaissance over Kilwa Kiswani and Kilwa Kevinge on 3 August.[27] *No changes of any description were observed.* Photos were taken of the road running South showing bridges intact. On the 6th No. 8 Squadron spotted for *Challenger, Talbot and Himalaya*, over Dar es Salaam. *Shooting on the whole was very good and considerable damage done.*[28]

The CinC, Cape, recorded that during the capture of Bagamoye by a naval force on 15 August, *Flight Lieut. Moon responded to my call for him with great promptitude and flew across from Zanzibar fully equipped with bombs, and photographic apparatus, and W/T. Having dropped his bombs round the inner trenches he took most excellent photographs, and then remained spotting for the big ships until fire ceased at 9 1.m. He was ably assisted by Air Mechanic Wilmshurst.*[29]

Gallehawk noted that a *considerable amount of flying was carried out here in 150 (Sunbeam) short* (sic) *seaplanes., which were well suited to the climate. One machine at least was usually away in some ship on patrol, or on reconnaissance duty over Tanga and Dar-es-Salaam.*[30]

Between the 21 August and 22 September, Flt Lt MacLean and Sub-Lt Fitzherbert in No. 3096, carried out a number of reconnaissance and bomb-dropping flights in the vicinity of Dar es Salaam and the Rufiji delta. Enemy trenches, gun positions, etc., were photographed and later bombed.[31]

Cull left on 9 September 1916, with Gallehawk and 34 ratings, the original members of No. 4 Squadron, to return to England.[32] Before then, he had taken the Khalifah of Zanzibar for a flight.

I was invited by his Excellency Flight Commander G.T. Cull to visit him at the aeroplane station. I went there and he received me with due honour and treated me with cheerful affability. He explained to me the science and skill required for flying and gliding in the air at will and discovering the enemy's hiding places; also about the heavy destructive bombs and deadly balls of fire used to bombard the enemy from above. I availed myself of the invitation to go for a flight and entered the aeroplane with him. It took us up a great height, rising and descending in perfect safety, thanks to the expert handling of the machine.

When he (the Flight Commander) goes to his native country, I pray that God will preserve him and enable him to perform good service. I tender him abundant thanks and high commendation, for he has often discomforted the enemy by sudden raids besides rendering conspicuous service to the fleet, and has shown great ability in the discharge of his duty. May prosperity and safety attend him whether he goes.[33]

Cull was replaced by Sqn Cdr Fredrick William Bowhill who transferred from Mesopotamia. Lincoln recorded in his

Above: No. 3096 "Britons Overseas No.4," being erected by RNAS Squadron No. 8, Chukwani Bay, Zanzibar.

Above: No. 8649 being erected on the wharf at Kilwa, East Africa by No. 8 Squadron, RNAS, in June 1917. Note the lack of tailplane, the empty packing case, and the box belonging to No. 8609. Another Sunbeam built 827 Type of the batch Nos. 8630 to 8649, this example has tropical side radiators. No. 8649 was carried by HMS *City of Oxford* in June 1918.

Above & Below: The Sultan of Zanzibar and Flt Cdr Cull watch as Lt Reg Moon takes No.3096 out on a demonstration flight for the Sultan.

Above: All hands to No. 3098, "Britons Overseas No. 8". There are at least nine men in the water to handle the seaplane. There must have been a fair wind blowing as the airscrew is not moving and the ailerons have moved about half way up from the at rest position. The different bathing attire, or lack of same, is noteworthy!

Above: *Self & G* off reef. Another photograph from Cull's album.

Above & Below: Bringing the seaplane ashore with the help of enthusiastic locals. The long flat beach made the launching and recovery of the Shorts a hard task.

Above: This No. 8 Naval Squadron Short 827 has the side panels removed to cool the engine. The flat object attached to the fuselage next to the front cockpit is thought to be an auxiliary oil tank. Note the man holding the starboard aileron up. The ailerons were not connected and dropped down when the machine was not underway.

Below: A Short 827 on board what appears to be a merchant ship. Probably from Zanzibar, 1917. The front engine cowl has been opened up to provide more cooling air to the engine.

Working on the 150-hp Sunbeam engine of one of the Short 827 seaplanes.

Below: This Short 827 is on a ship. Note the light (grey?) engine panels. The wing fold attachment is well displayed.

Above & Below: Moving No. 3098 into a locally built shelter at Kokotoni, East Africa, 1916.

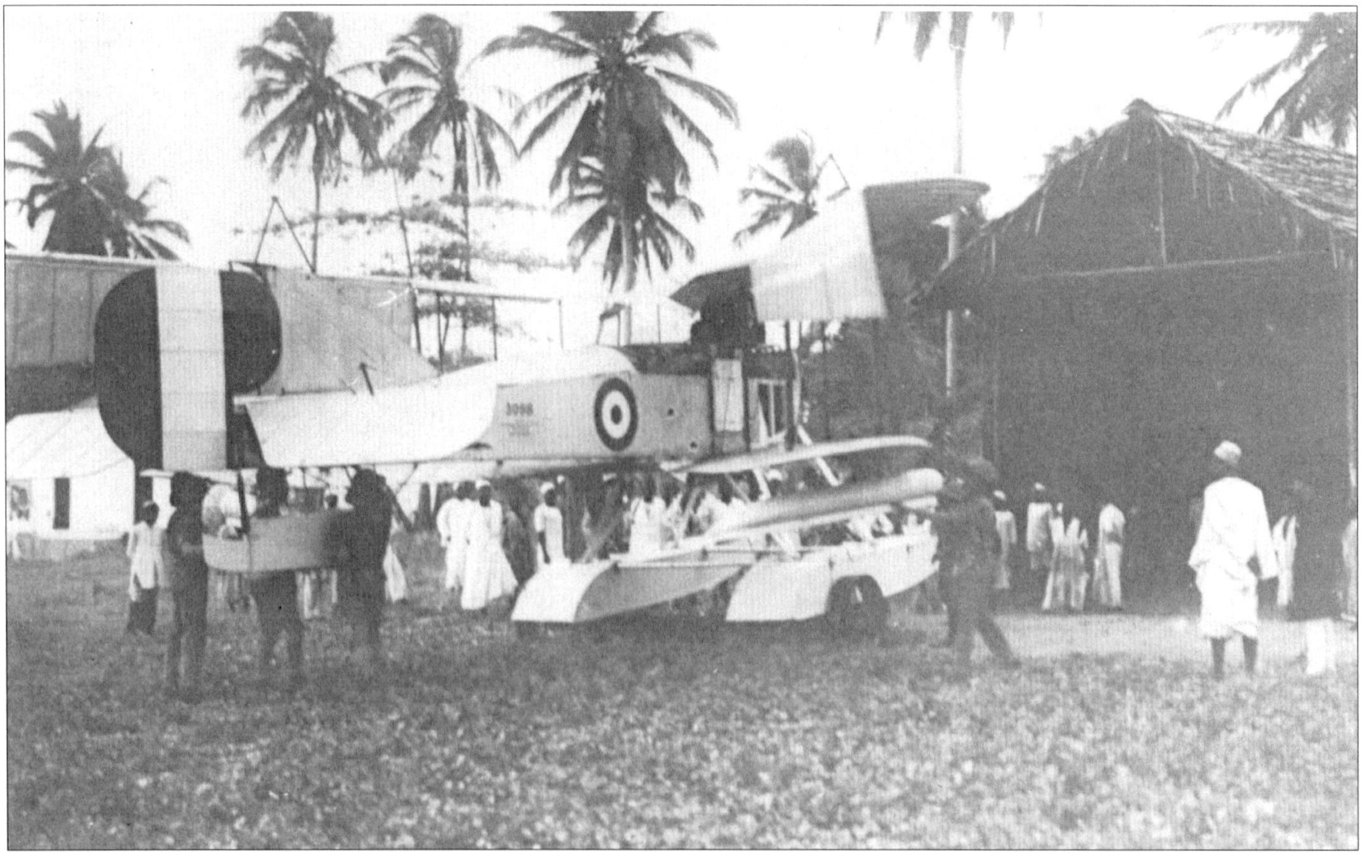

Right: Short 827 of No. 8 Naval Squadron being hoisted on HMS *Himalaya* off Dar-es-Salaam, 6 August 1916. Noteworthy is the front engine cowl panel that has different cooling holes when compared with others of the type.

Above: No. 822 with a land undercarriage. Note the Union flag is now on the fin. The engine panels have been removed due to the climate. First of the batch of six (Nos. 822 – 827), Sent to RFC Force D in Mesopotamia, arriving on 5 September 1915, it was erected as a seaplane. The land chassis was fitted in mid-November 1915. It crashed 10 miles south of Kut on 14 February 1916 and written off.

Left: Details of the bomb carrier on a Short 827 Type with what appear to be two 112-lb bombs loaded. Isle of Grain, 6 June 1916.

diary that seaplane work *has been standing by ever since the surrender of Dar es Salaam*. On 4 September, the land forces entering the town to find that the Germans had abandoned the site after destroying major facilities likely to be of use to the British. Lincoln continued: -

This is not to convey the impression that little or no work has been performed. On the contrary, ever since the arrival of Cdr Bowhill, the squadron has been if anything working at higher pressure than during any previous period. In the first place, the patrol ships have been extra busy during these three remaining months of the year and our seaplanes aboard them have been carrying out the duties of aeroplanes, flying many miles inland and strafing enemy troops and encampments with bombs, high explosives and incendiaries. Again new aerodromes have been opened up, this, of course, involved a further disintegration of the squadron and to make matters worse most of the men sent have gone down with malaria and have repeatedly to be reinforced from the main body at Chukwani. Furthermore, much work has been entailed in almost wholly renovating the original seaplanes and, at times, owing to the loss of fresh stores, engine parts, etc, vast quantities of which went down on the SS Mombasa, the squadron has been hard put to it to maintain a sufficient number of well-used planes in good flying condition.[34]

It is not exaggerating to say that our seaplanes have gone a very long way indeed to win the campaign in German East Africa, for the enemy were undoubtedly endeavouring to hold the coast line with as strong forces as circumstances would admit. Most of their big guns were in position for coastal defence. It is true they made a poor show of resistance but the bombs from our machines took all the heart out of them.[35]

One such flight was carried out on 11 November by Flt Lt Moon with AM F Wilmshurst in No. 8642. The machine carried two hours fuel, two 65-lb bombs, W/T and a camera. The object was a bombing and reconnaissance flight over Mrweka (Lindi). The flight left at 0600. Over Mrweka one bomb was dropped on a house near the quay. The second bomb was dropped on another house and was seen to explode a little left of the target. A W/T message, to the effect that no movement could be seen, was received by the armed whaler *Salamander* that was then observed to put about. They returned to the ship at 0750. Photographs were taken of the creeks and rivers, probably to be used to update maps of the area.[36]

The December 1916 Report of No. 8 Squadron noted that since August *Short seaplanes have been flying continuously, and have proved to be of great value – they stand the climate well, and also the strain of being continually hoisted in and out of ships. The type also behaves well in the "bumpy" conditions experienced in this atmosphere. 150-h.p. Sunbeam engines have flown 150 hours without a breakdown – with one exception – a very credible performance considering the climate.*[37]

No. 8254 was lost while conducting a reconnaissance of the Rufiji Delta on 6 January 1917.[38] The seaplane from *Manica* was unavailable as its fabric was so slack it was unable to carry an observer. It was being returned to Zanzibar for a complete overhaul. No. 8254 left the *Himalaya.* to carry out the reconnaissance with Flt Lt Moonas pilot and Flag Cmdr the Hon R.O. Bridgeman, DSO, as observer. Bridgeman wanted to determine whether the small steamer that had been observed two days previously was still in the river. If so, he was going to send armed Motor Boats up the river to try and cut the steamer out. Moon later wrote that it *has always been my practice to wait until I return from a flight before taking a meal, because I believe that work of his kind is better done on an empty stomach, and so I had nothing more than a cup of tea before leaving. If I had known what was in store for me, I might have been tempted to stow enough inside me to last me a week, after the manner of a pelican.*

We made a very through reconnaissance of the delta, flying over all parts of it, and at the end of an hour or so the commander said he was quite satisfied and ordered me to return. We were over the south end of the delta when the engine revolutions suddenly began to drop, so that I was obliged to descend. I steered for the inshore end of the Suninga Channel, and landed in a creek which forms a junction between this and Kiomboni Channel to the south of it. I taxied along the creek, while the commander took the pilot's seat to give me an opportunity of attending to the motor.

I found that the after magneto drive had failed, and presently the pressure in the petrol tank gave out, and the engine stopped altogether. I made several attempts to restart it, but without success. The only way to discover what was wrong with it was to take it to pieces, but of course I had no idea whether I should be able to repair the defect when I found it. The commander decided that under the circumstance there was nothing for it but to destroy the seaplane, and try to make our escape to the south of the river....

The first thing we had to do was destroy our seaplane, which we did by soaking it in petrol and setting fire to it with a Verey's light. We watched it burn until it was a useless wreck.[39]

The seaplane carried fuel for about 3 hours 24 minutes. When it failed to return it was decided to send up another seaplane to look for the missing machine. Flt Lt Sub-Lt Deans took No. 8253 up at 1050 followed by the *Himalaya* to pick it up and the seaplane returned in about an hour. Deans had been told to check out a column of smoke that turned out to be a bush fire. He then had to return for more fuel. The floatplane was refuelled and left again at 1225. This time Deans was solo as the machine would not carry an observer and a full load of fuel. Deans reported that he had searched *Arms of Delta to the South and all creeks and channels, between and around Kikikela, Kikmiya, Mbunu, and Suini. When returning by Sasiba Uranga arm* (sic) *observed Seaplane in water by left bank 6 or 7 miles from mouth of arm. Landed machine alongside, and found that the machine in water was completely wrecked. It was lying about 10 yards from the bank, upside down with the bottom of floats above water, and portions of wings, tail and rudder showing, which were burnt. From the position of the wreck it seemed probable that it had nose dived into water and turned over, also that it was on fire, as the fabric under the water as far as could be seen was not burnt. It was lying in about 8 to 10 feet of water. Tide was then high. After waiting, rose and returned to H.M.S. "Himalaya."*[40]

On the 7th, Short No. 8642, with Smethurst as pilot, was to fly to Dar es Salaam to help in the search but came down about 10 miles from Chukwani. The crew spent 5½ hours clinging to the wrecked seaplane's floats before being picked up.[41]

In addition to the seaplane the ships *Thistle* and *Echo* were to search the northern mouths and when the tide permitted, were to go up the channels. *Thistle* proceeded to within two miles of the wreck before shoal water prevented it from proceeding further up the river, but it remained three miles up the river all night. *Childers* was sent to the Msala mouth channel. *Mancia* remained at Niororo, in case the seaplane returned there, it being the rendezvous. It was considered that both pilot and observer must have been instantaneously killed.

On 24 January 1917, GHQ, Dar es Salaam, was telegraphed that a prisoner had informed that Moon was a PoW but that Bridgeman had died. A Court of Inquiry was held into the incident the following month. Evidence was presented that the machine was new and a good one. Reconnaissance flights were usually carried out at about 2,000 feet. The machines had been fired on and hit on previous occasions. The report of the Court was of the opinion that the machine had been shot down or had to come down due to engine trouble. The officers themselves had set the machine on fire before attempting to escape. *Everything possible was done by those on the spot, and it was especially fortunate that the only other improvised seaplane ship on the Station was available and able to ascertain to once there the missing seaplane had descended.*[42] The Operations Report of the RNAS of April 1917, recorded that for Squadron No. 8, based at Chukwani, Zanzibar: -

January 6th. Accident. -Seaplane No. 8254 was lost whilst flying over the Rufigi Delta. The accident was probably due to enemy gunfire, and it is not unlikely that the machine caught fire in the air.[43]

Capt Schade of the German steamer *Markgraf* was at the time of the seaplane incident in the Rufigi Delta and afterwards was in charge of Flt Lt Moon for several weeks. His story, that was a mixture of what he saw, was told by Germans and natives and by Moon, is as follows: -

On January 6th, Captain Schade, and a few Askaris, was doing duty at Mbumi. Early in the morning they heard and saw a Seaplane over the Rufigi and evidently making an Observation over the Konigsberg. While they were watching she turned as if to fly towards the Mbumi River and they heard what he described as a change in the noise made by the Seaplane engine as it had stopped and it was evident that the machine was coming down to land on the water. Owing to the bush he could not see where she landed but there was no indication of a fall; the machine merely made a slant downwards as it making an ordinary landing. … It had not been fired at on this flight but was forced to land owing to magnetic failure.

Once landed, and having ascertained that the damage was not repairable, the two officers proceeded to burn the machine. … they at once proceeded to make their way up the bank of the River towards Betya. Anything but the slowest progress was impossible, and it was not until the afternoon that they reached to opposite Somali.

Moon swam the river to see if he could find some food and possibly a boat. Captain Schade mentioned his admiration for the pluck of Moon in making the attempt as the river was swarming with crocodiles. The two officers were separated and on the following morning recommenced heading towards Betya. Schade was of the opinion they kept in touch by firing their revolvers. Moon swam back across the river but did not take into account the strong ebb tide and ended up some distance downstream and the two did not meet up until late in the day.

On the night of the 7th they found an empty house and constructed a raft using a window frame as the base and roughly fastening planks across it.

While making their way on foot Moon told him that they were frequently up to their necks in water, and not only that but the only relief they could obtain from the millions of mosquitoes was to continually dip their heads under the water.

During the night of the 8th they set off again on their raft. Bridgeman was practically insensible from this time onwards. At Kiomboni the raft was carried out to sea through the Simba Uranga Mouth. On the last day, the 9th, Bridgeman finally died of exhaustion and exposure or he was washed off the raft. The tide brought the raft back close to shore and Moon struggled to shore. When he was finally on land his hands and feet had been cut to pieces by the rocks. A native found him and took him to a house where two Germans were living. They fed and clothed him and when he was sufficiently recovered took him to Msala where he was placed in the charge of Capt Schade. Bridgeman's body was washed ashore a few days afterwards and buried by the Germans.[44]

Moon was released after 10 months imprisonment and then served in the UK; and he corroborated Schade's statement.[45]

Following the taking of Dar es Salaam, work for the seaplanes became less. The Voisin aeroplanes that had been sent out were uncrated and were able to operate successfully inland in support of the Army.

The seaplanes still continued to undertake work. No. 8253 under Flt Sub-Lt C.F.M. Chambers with PO (M) F Wilmshurst flew from *Manica,* and after her return to the UK, from Lindi harbour, mapping the interior.

In May No. 8 Squadron carried out 34 flights over enemy positions and camps near Lindi. A number of bombs were dropped and 2,000 darts. Seaplanes also carried out patrols from Zanzibar to search for signs of hostile craft. The following report of the work done by No. 8253 gives an idea of the work of No. 8 Squadron's seaplanes through this period:

03.05.17. One 65-lb dropped on red-roofed building at end

of Mayani, but missed target. Darts were dropped as well as an incendiary bomb, but results could not be observed.
05.05.17. Two 65-lb bombs dropped on trenches in the Mingoyo district but results were not observed.
08.05.17. One 65-lb bomb dropped on main village, Mingoyo, but results not stated.
09.05.17. One 65-lb bomb dropped on a camp near Schaeffers Farm & 1,000 darts on another camp W of the farm, but results not stated.
13.05.17. One 65-lb bomb was dropped on an Arab house in Mongoyo, another on the salt works 1 ml from the pier on the east side of railway running from Mkwaya to the Präsident, and started some water running.
18.05.17. One 65-lb was dropped on some huts to the west of Schaffers Farm, falling close to the huts.
21.05.17. One 65-lb bomb was dropped at the 4.1-inch gun position at Mingyo, but fell 100 yds short. Another 65-lb was dropped at Schaffers Farm but fell 50 yds short of the house.
25.05.17. Two 65-lb bombs were dropped at Schaffers Farm causing some damage to the building. Shortly after a second attack was made and one 65-lb bomb was dropped at the farm which hit the NE side of the buildings. Another 65-lb bomb was dropped on a building in which a number of men had taken shelter and a direct hit was scored.
26.05.17. Two 65-lb bombs were dropped at Schaffers Farm, one falling 30 yds short of the building, causing apparently very little damage. A series of photographs of the section of the river from Mkwaya to the President were taken.
30.05.17. While on a reconnaissance flight, one 65-lb bomb and two incendiaries were dropped at Schaffers Farm. The incendiaries failed to ignite possibly due to their falling on sand and becoming smothered.
31.05.17. Two 65-lb and two incendiary bombs were dropped at Mkwaya, and although he bombs dropped amongst houses no damage seems to have been caused. Note on two occasions the General Commanding the District was taken up as a passenger for reconnaissance purposes.[46]

The crew during this period was Flt Sub-Lt Chambers, with his observer, F Wilmhurst, PO Mechanic (W/T). Brig-Gen O'Grady called attention to the pair who were willing to fly without regard to their safety, carrying out reconnaissance and bombing flights, assisting the Survey Department in map-making and carrying despatches when other methods of communication between the Army columns failed. They were *always ready to fly at any time and any weather.*[47]

The use of darts is noteworthy. The Shorts were adapted locally for bomb dropping and the reports show that the aircrew's expertise was not good. Given the condition that the Squadron operated under, the seaplanes gave a good account of themselves.

From 10 to 13 June 1917, the British carried out attacks to secure Lindi and capture the Germans 4.1-inch gun that *lately had been giving a great deal of trouble.*[48] The operation was successful but the 4.1-inch gun was not captured.

With the conflict heading inland the seaplanes were withdrawn to Zanzibar for overhaul. Four surviving Short 827 seaplanes were reportedly given to the Belgian African forces. The late historian Jack Bruce noted that if these machines were the same ones used by No. 8 (Naval) Squadron *their serviceability record must have been unequalled by any contemporary seaplane.*[49]

Lincoln summed up the squadron's first half-year's worth of flying as follows:

It is interesting and at the same time immensely gratifying to review the first half year's work accomplished by this squadron. In a way we have been pioneers of flying in this particular region. When we first arrived we were treated to stories of the immense, almost insuperable, difficulties of flying out here. It was rumoured that aeroplanes experienced the greatest difficulties in rising from the ground and that when in the air it required a super-pilot to manage them; how could seaplanes with their great heavy floats hope to put up a satisfactory show?

He then attacked the efforts of Cdr Cull and his men. Lincoln's criticism is not fair as Cull's problems are well documented. Lincoln continued: -

We arrived at Zanzibar midway through March and, of course, a reasonably lengthy period elapsed before matters were sufficiently well in hand to permit of our pilots getting to work. The hangar had to be got ready, the ground levelled and the slipway constructed. All this took time; it was not until the end of the month that our seaplanes were seen on the wing when the hangar was ready for their reception. Lt Moore made all the test flights with conspicuous success. Dar es Salaam was the first place to which our airmen turned their attention in accordance with orders received from the Admiral. Cdr Nanson and Lt Moore made most of the initial flights taking with them observers who obtained a great deal of most useful photographic information It appears that during these early days of operations very little was known of the fortified conditions of Dar es Salaam. Since the latter part of April until its surrender on 3 September (it) has been subject to more or less heavy, intermittent bombardments by our ships.

When Cdr Cull and his crowd came here early in May, Cdr Nanson with a few of our men had already gone up-country to Mbuyuni as a relief party. Since then, the original party has been considerably reinforced from our squadron and this little offshoot of the main squadron has done some magnificent work with land machines, following up the army. In addition to machines flying from the station to various scenes of operations, we are still able to send one or two machines with pilots on patrol ships and these do much useful work in the shape of daily flights from the boats over various coastal towns, bomb dropping and photograph taking. More than once has our squadron been complimented by the Admiral on the splendid work done. Whenever and wherever our planes have been wanted they have never failed to appear and that is saying a great deal where seaplanes are concerned. A machine becomes greatly enervated after a long flight of 2 or 3 hours duration and an overhaul of a more or less complete nature is essential after each trip. Finally, it may be said that our flying squadron has been made responsible for the patrolling of a most extensive strip of coastline extending from Tanga in the north to Lindi in the south.[50]

The official 'Report on the East African campaign' mentions the Short 227 Type 150-hp Sunbeam. This is thought to be in error for the 827 Type. It described the machine as follows:

This seaplane was used in East Africa, and proved to be satisfactory in every way for the work required, namely, reconnaissance, spotting and photographic work with the occasional bombing flights. This seaplane stood up to the climate well, and also the continual hoisting in and out of ships. In spite of the heavy heat bumps experienced during the day, this type of seaplane was generally controllable, though it was very hard work to fly them over any length of time. By constant care, overhauling, recovering, etc. two of the seaplanes which left England in January 1916, are still on the active list for flying. (Feb. 1918).

As these machines had to work away from base, all were altered to hand starting, so as to avoid the possible delay of having to wait for air bottles to be recharged. This method proved, in every way, to be satisfactory, and in my opinion, should be employed whenever possible on machines detailed for stations which are not easily accessible.

It must be borne in mind that in German East Africa we had no hostile aircraft to contend with, and the anti-aircraft guns were very bad. If this had not been the case, a faster and better climbing machine would have been required.[51]

The author of the report, thought to be Fredrick Bowhill, then set out what he considered the best machine for the tropics stressing the use of metal wherever possible, the machine to be able to be broken down into small parts capable of being carried by porters.

Mesopotamia

For an overview of the Mesopotamian campaign and the use of air power, Volume V of the Official History, *The War in the Air*, covers this, with the usual omissions of aircraft designations, serials and the first names of personnel, however, it is still the best coverage of the campaign.[52] The first Battle of Kut involved the Turks besieging the British Army garrison in the town of Kut-al-Amara. Gen Townshend surrendered and about 13,000 Allied soldiers went into captivity. This defeat was the worst defeat of the Allies in WWI.[53]

As noted in the Short 166 Chapter, three 827 Type seaplanes were sent to Mombasa in July 1915, to help in the campaign to destroy the German raider *Königsberg*. On arrival they were sent instead to Mesopotamia. On receiving instructions to dispatch a *flight of naval seaplanes for work in Mesopotamia, the Air Department detailed four officers, including the Officer in command, Squadron-Commander* (Robert) *Gordon, one warrant officer, and twenty-one ratings for this duty.*

Three Short seaplanes (150 h.p. Sunbeam engines), and one Sopwith seaplane (100 h.p. Gnome engine) were sent out to start the seaplane station at Basra.[54]

Flt Lt Alan George Bishop, who we met above, was attached to the RNAS party that went to Mesopotamia. At Bombay they received four 150-hp Sunbeam Short seaplanes. Short Admiralty 827 Type Nos. 822, 825 and 827, with 150-hp Sunbeam engines, are known to have gone to Force D in Mesopotamia. On arrival at Basra the machines were to be assembled as quickly as possible and to join up with the Army at Kut-al-Amara.

A large iron barge was taken over, the top deck being converted into a hangar, the remaining space being fitted out as workshops, stores and living quarters. This barge was most useful and always accompanied us up stream.

On arrival at KUT we found we were in time to take part in the operations leading up to the battle of Ctesiphon and the subsequent disastrous retreat. The reasons for this failure are now a matter of history, and I do not propose to touch upon them.

During this period we were employed on artillery co-operation and reconnaissance work. As regards artillery co-operation, I do not think the army placed much value on our services. This was chiefly due to the fact that the country being so flat, there were few targets which they were unable to see.

After the retreat and the investment of KUT, we were employed in reconnaissance and bombing operations. Later on when the garrison at KUT became short of supplies, all aircraft were employed in dropping food, money and miscellaneous stores in to the town. Although this assistance enabled the troops to hold on for a few more days, disease eventually compelled them to surrender.[55]

According to Bishop the Cin C has asked for seaplanes for Mesopotamia *because of the heavy floods at the time of application.*

Actually by the time we arrived, the floods had subsided and it was soon discovered how unsuitable seaplanes were for the country.

Apart from the bad moral effect of flying in a seaplane over the desert, there was (sic) also other difficulties to be overcome.

Firstly, the river current was so strong, that when the seaplane was fully loaded, unless the wind was favourable, it could not get off the water. This was also partially due to the heat opening up the seams of the floats and allowing water to enter. Secondly, the engine was started by compressed air, and the nearest charging station for the air bottles was H.M.S. ALERT *and* ABADAN.

In view of the above two land undercarriages were made, and two seaplanes were converted; also hand starting gear was fitted to their engines. Both aircraft in their new condition worked splendidly.

By this time, the mistake in sending seaplanes had also been realised at home, and both R.F.C. and R.N.A.S. detachments equipped with B.E.2C'S, VOISINS, and HENRI FARMANS were beginning to arrive in the country.[56]

Seaplanes Nos. 825 and 827 had been tested on 25 September 1915. No. 822 was still on the barge following up the river. The two left the Sanyat camp flew up to the Nakhailat advanced camp on the 26th, and although heavy guns were firing at the Turkish positions, the seaplanes were not required to spot. The following day No. 827 carried out a successful W/T test but was not required to spot. Finally, on the 28th, Lt Blackburn and Arnold in No. 827, carried out spotting for the five-inch guns. The W/T worked and all signals sent were received by the guns. The flight lasted from 0530 to 0635 when the seaplane was caused to retire owing to a defective petrol line. The leads from the gravity tanks to the engine on all three machines were *very badly placed and caused air locks. They were all altered.*

The OC of the artillery informed the RNAS that once the guns were ranged in there was no point in the machines remaining on station, however the observers reported that *although the guns received the W/T signals from the seaplanes correctly they paid no attention to them, unless they agreed with the spotting of the forward Observing Officer of the Battery, and the firing was not as good as it might have been if the guns paid more attention to the seaplane observer. The target was plainly visible, and flights were made from between 2000 & 3000 ft at a range of 6500 yds.*

On the 29th Nos. 825 and 827 both carried out a successful reconnaissance, the enemy being located in full retreat. On return, No. 827's *engine failed and on landing she was blown on to a lee shore where the elevator and ailerons were badly damaged putting the machine out of action.*

On the 30th, stores and lighter arrived from Nakhailat. No. 827 had to be left behind under guard as there was no transport available for it. The RNAS took over an island for a seaplane station. On 1 October 1915, Maj Robert Gordon and Lt Bishop left at 0625 for a reconnaissance flight, but had to return due to the sickness of the pilot. *Both pilots were suffering at the time from fever, and the heat of the machine owing to the slip stream blowing directly through the radiator on to the pilot's face, makes these machines very trying to fly in a hot climate.* An attempt to tow No. 827 failed when the tug ran aground and returned without the seaplane.

The 2nd saw No. 825 start a reconnaissance at 0545 by Maj R Gordon who was solo with six 20-lb bombs. He had to return due to petrol failure. At 0730 he started again but hit a native boat on take-off, severely damaging the chassis, causing him to return to base. *This accident was caused solely by the pilot not being in a fit condition to fly.* Both machines out of action and no tug available to tow No. 827 up from Nakhailat. Lt Arnold accomplished this the next day and returned at 1630 with No. 827 and repairs commenced immediately.

At 0735 on 5th, No. 827 left for a reconnaissance with a solo Maj R Gordon. He was to reconnoitre Tesiphon where the enemy was supposed to be. At Aziziya the airscrew sheathing stripped and vibrations broke the fuel pipe. He landed and made repairs; the airscrew later being repaired with solder. Continuing the flight at 1210 he found the enemy's trenches astride the river at Ctesiphon but no sight of troops. He returned to Kut-al-Amara at 1625 after flying some 240 miles. Orders were received from the Army Commander on the 6th that owing to the lack of compressed air and the non-arrival of the seaplane barge, *no more flying was to be done by the R N A S until further orders.* No. 825 was hoisted onto trestles and a land chassis was commenced to be fitted. There were no spare air wheels with the two land chassis. As the machines would be more useful on wheels and only two chassis were received and no spares, *I am only putting one of them on wheels.*

Dual ignition was fitted to No 825 in order to start it by means of a starting magneto. It was not possible to start the machine by hand. Only two filled air bottles for the starters were available. Six more were on their way after being filled by HMS *Juno*, and *have been a month on their way, and have not yet reached us.*

Sickness led to a working party of nine soldiers being permanently attached to the RNAS detachment, but they were untrained. *At the moment I have only one carpenter R N A S and one lent from the R F C for five days.*

The machines themselves are very fast but will not get above 2000 ft with a passenger and full load. My orders are not to take a passenger for reconnaissance, but as a single seater the view is bad. The enemy have some good Krupp guns but make bad shooting, the machines only being touched twice so far.[57]

According to Sturtivant and Page, Short Nos. 822, 825 and 827 were the two seaplanes given a land undercarriage in October 1915.[58] They were used against the Turkish forces advancing on Kut-al-Amara. The British Official History states that the two were converted in October and *although the Sunbeam engines gave trouble, the Shorts did good work.*[59] No. 827 was reconverted back to a seaplane after being unserviceable by December 1915. It was tested in its old configuration in January 1916, and survived until deleted on 1 January 1917. No. 825 's fuselage was too badly warped for repair by 6 December 1915, but was reconverted to a seaplane by April-May 1916. This appears unlikely unless a new fuselage was available. It was deleted in 1916.

Several Short 184 seaplanes were sent out before it was realised that they were not suitable. Nos. 8043 to 8047 are known to have been sent to Mesopotamia. They took part in the airlift of supplies to Kut. Conditions were harsh and the seaplanes suffered as reported: -

All seaplanes have been kept in repair as much as possible. These machines take a lot of hands to handle them and often have to be attended to day and night as the winds during the last month have been very strong at times, and sudden shifts are frequent.

Considering that they are entirely exposed to all weathers, the damage they have sustained has not been very great but nevertheless sufficient to keep the crews and repair parties busy at them all the time.

Also the mud here is very tenacious and getting machines on and off requires a lot of handling and pushing, and throws them out of truth.

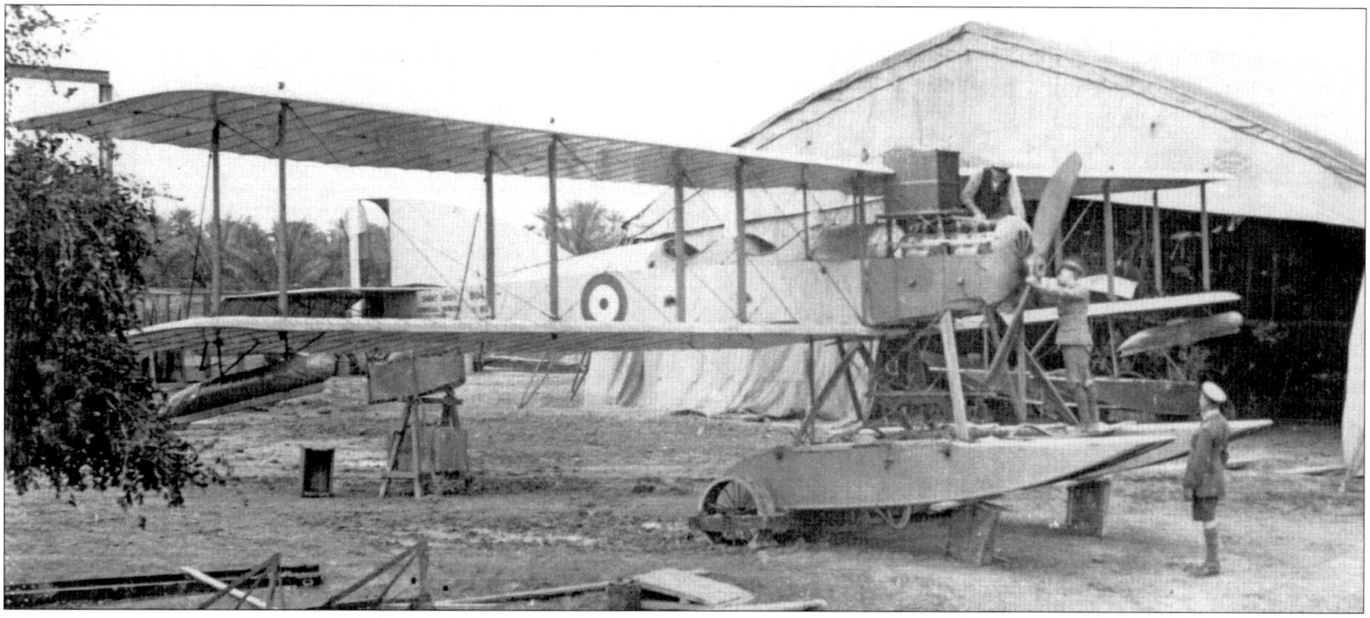

Above & Top: No. 8043 (S.185) was a Short manufactured 225 Mohawk powered 184 Type from a batch of 75 (Nos. 8031 to 8105). It was shipped to Mesopotamia in the SS *Hunts Castle*, and is shown at the RNAS main base at Shatt al-Arab, Basra. It arrived before April 1916, when it was recorded dropping supplies into Kut. It went to Port Said from August and had returned to the UK, being recorded at the Central Supply Depot White City, by 31 December 1916, and was deleted the following month.

Above: Short 184 No. 804X. Note the unusual carrier mounted under the fuselage, possibly that used for the supply of Kut, during which action both machines performed supply drops.

Below: The RNAS main base in Mesopotamia was in a building on the banks of the Shatt-al-Arab at Basra. The building was formerly a Turkish hospital. The Bessonneau hangar belonged to Maj R Gordon's Squadron and was erected on 14 December 1915, after the units escape from Kut-al-Amara and the besieging of the town. Short No. 827 is just visible.

The report also noted that a fair amount of sickness had been experienced for the month with seven officers out of action for 14 days and 30 ratings for 26 days.[60] Aubey Herbert wrote of the flies in Mesopotamia that *They hatched out until they were almost the air. They were in myriads. The horses were half mad. The flies were mostly tiny. They rolled up in little balls when one passed one's hand across ones sweating face. They were on your eyelids and lashes and in your lips and nostrils. We could not speak for them and could hardly see.* Murray Sueter quoted this passage from *Mons, Anzac and Kut,* (The Adjutant's Press, UK, 1919), to illustrate the problems with truing up an aeroplane in these conditions.[61]

On 26 April 1916, Flt Lt Cecil B Gasson with 2/Lt Archibald Cecil Thouless were shot down while dropping supplies into Kut from Short 184 No. 8044. Gasson was wounded and Thouless killed. *The attacking enemy machine was a Morane type of monoplane believed a Fokker fitted with machine-gun which fired through the propeller. This machine attacked ours from behind and then climbed rapidly over our machine afterwards returning & attacking again from behind. It used armour piercing bullets of hardened steel, with a diameter of 6.1 mm & length 33 mm. The gun was not rifled.*[62] Gasson was made a PoW. The victor was *Hpt* H Schüz of *FAb* 2.[63]

The RNAS lost another aircraft that day when a B.E.2c on supply dropping duty was shot up, the right aileron being shot away. The pilot, Lt Donald Davidson, was severely wounded. He had 32 bullet holes in his machine.

Albert E Cowton who was serving at the time with an RNAS Balloon Section wrote that before the fall of Kut a Short seaplane had been used, *operating from the river, and it had given a good account of itself. It fell into disuse, however, owing to the difficulty of obtaining spare parts.* It was dismantled and the floats were handed over to the Section. As they did not have the materials and labour to make a raft from them, the CO allowed some of the local men to make use of them for fishing. *To these people, life afloat is second nature.*[64]

The surviving Short floatplanes were withdrawn to Basra in December 1915, for a refit. It was decided to form a composite flight of RNAS and RFC personnel at Ora, and the Shorts were flown there. One 827 Type was wrecked on 31 January, a second was damaged in a bad landing on 4 February, and No. 82 on the 14th turned over on landing and was wrecked.[65]

An official report noted that the *Short Seaplane 225 HP Sunbeam* was used in Mesopotamia, *and was found to be slow in climbing, taking from 60 to 80 minutes to reach a height of 5,000 feet with pilot, observer, Lewis Gun, W/T, three hours oil and petrol. This seaplane was very difficult to control in bad heat bumps, but, at the same time, never quite got out of control.*

For working on the Tigris, a comparatively narrow, tortuous, and fast running river, this seaplane was not suitable on account of her weight, and the length of run required to get off the water. A lighter type of seaplane would have been more suitable... a boat type would have been best.

This seaplane stood the rough handling required to get her on and off the tenacious mud banks in an extraordinary manner. (Sometimes I had to use luff upon luff to move them). No other seaplane type, fitted with floats, would have stood halfa the treatment these seaplanes experienced, and continue to fly as they did. It speaks highly of their good and strong construction. (These were built by Messrs. Short Brothers, Rochester). [66]*Ailerons and elevators were liable to warp in the intense heat, and should be strengthened to prevent this. The longerons also warped.*

Home Waters

In Home waters the 827 Type performed the usual boring and dangerous patrols from 1915 till the Armistice. A few

Summary of Work for April 1916

Flying Times			**Made up as follows:**	
Type	**Serial**	**hrs-min**		**hrs-min**
Voisin	8506	25-47	Spotting	35-05
Henry Farman	1450	41-25	Food Carrying	89-50
Short 184	8043	20-00	Reconnaissance	15-10
Short 184	8044	18-05	Escort	1-30
Short 184	8046	36-34	General	16-06
Short 184	8047	15-50		
	Total	157-41		157-41

The Total amount of food dropped into Kut is as follows:
Aeroplanes: 5,100 lbs
Seaplanes: 3,290 lbs
Number of flights: 47
Also, some 500 lbs dropped in RFC machines by our pilots.
Source: FAA Museum File: 1997/234/0003.

Above: Short No. 822, the first of the batch of six (Nos. 822 – 827). Sent to RFC Force D in Mesopotamia, arriving on 5 September 1915, it was erected as a seaplane.

Above & Below Left: Short No. 827 was erected on 8 September 1915 and flown as a seaplane on the 10th. Converted to a landplane in October, it flew in support of the 6th Indian Division until Kut-al-Amara was besieged.

Above: photograph of Short No. 827 on the Tigris River, November 1915. The employment of these seaplanes was due to the War Office not appreciating the actual conditions in Mesopotamia.

incidents involving the type are recorded below:

Short 827 No. 8221 underwent its one-hour acceptance test on 1 February 1916 under the hands of Sydney Pickles. Seven days later the machine left Yarmouth for Killingholme under Flt Sub-Lt Eure. At 1415 NE of Winthorp with the prop slowly working heading toward shore. At 1430 the engine stopped for 10 mins and she drifted NE out to sea. She was then about three miles away. A boat was launched from Skegness Coastguard Station but did nothing. The seaplane had alighted at 1440. At 1845 a report was received from Sutton-on-Sea that four Verys Lights were fired SE of Anderly Coast Guard Station. Skegness launched a lifeboat; Immingham Depot sent a destroyer and patrols in neighbourhood were instructed to go to assistance. However, they searched in vain. On the 9th the *SS Blonde* sighted No. 8221 in distress at 0730 with Eure on board. He had been in the water 19 hours. The seaplane was taken in tow and while off Yarmouth transferred to Motor Patrol Vessel *Platypus No.4*.[67]

On 15 February 1916, Flt S-Lt Bernard Richard Lee and Flt S-Lt Herbert Joseph Page left Great Yarmouth in No. 8220 for patrol and never returned. Search for the missing men and their machine took place over several days but neither was seen again. Presumably they suffered engine failure and were forced to alight and thereafter were drowned as there was a heavy gale that night.

On 25 April 1916, when the German High Seas Fleet bombarded Lowestoft, a 827 Type from Yarmouth carried out a bomb attack on the German ships. This raid was planned as part of a substantial Zeppelin raid on London. The weather rendered the operation largely ineffective. Yarmouth sent up a number of aeroplanes, Bristol Scouts and B.E.2c biplanes as well as a solitary Short 827. No. 3105 with Flt Sub-Lt E Cadbury with Flt Sub-Lt A.H. Sandwell searched from 0340 to 0725 hours without success.[68]

The 28 November 1916 saw an unusual end for No. 3324. Flt Sub-Lt G.L. Davies, with Sub-Lt A.C. Stevens as observer, left Felixstowe on a reconnaissance flight. They never returned. Their machine had been forced down and the crew were picked up by a U-Boat becoming PoWs.[69]

On 20 December 1916, Flt Sub-Lt J.R. Ross in No. 3327 from Portland, sighted a submarine on the surface proceeding to attack a transport. The submarine submerged on sighting and pilot dropped one bomb but probably for no effect. The seaplane remained over the ship for 45 minutes until two patrol boats arrived, preventing the submarine from making a further attack.[70]

Dundee, 12 March 1917. Flt Sub-Lt McNicholl with LM Jerkins. A submarine was reported off Bell Rock in the afternoon. The sub was sighted about two miles NE of Bell Rock with a great part of the hull being above water, but she submerged before the pilot could get into position to attack. McNicholl continued to patrol the area and about 17.00 sighted the submarine six miles ESE of Bell Rock and

Left: Short No. 827 being serviced at the RNAS's main base at Basra, December 1915. Behind the Bessoneau hangar is the former Turkish hospital used as the RNAS HQ.

Right: *Under the Gallows.* Short No. 822 being serviced at the RANS Depot at Basra. It still has its floats.

Left: Short No. 822 at the same location with the floats removed and a tail skid fixed under the rear fuselage.

Right: Short No. 825 with its wings folded shows the land undercarriage. The engine panels were removed due to the hot climate.

Left: Short No 825 was tipped over on landing by Lt H.P. S. Clogstoun of No. 30 Squadron, with his observer, Lt W.G. Palmer.

Right & Following Page, Top: Short No. 825 of Maj R Gordon's RNAS Section with the combined RFC/RNAS Squadron in Mesopotamia. A Farman pusher is visible in the background; only second-rate machines were sent to this war zone.

Left: Short No 825 was tipped over on landing by Lt H.P. S. Clogstoun of No. 30 Squadron, with his observer, Lt W.G. Palmer.

Right: In this photograph of No. 825 being fitted with a wheel undercarriage, the position of the Union flag is well shown.

Left: View of the supplies for the besieged Kut-al-Arama that the Short 827 carried under the fuselage. Crah, February 1916.

Right & Below: This Short, No. 8043, was the usual mount of Flt Lt W.H. Elliott. It is preparing for a flight from Orah on the Tigris River carrying supplies to Kut-al-Amara in April 1916. The ship in the background is the monitor HMS *Stonefly*.

Right: One of the four Short 827 seaplanes that arrived with Cdr F.W. Bowhill in February 1916. The RNAS base on the Shatt-al-Arab is well shown with other seaplanes in the Bessoneau hangar.

Left & Below: No. 822 following its crash on 14 February 1916, when piloted by Flt Sub-Lt W.H. Dunn at Orah. Shipped backdown river to Basra it arrived at the RNAS base ten days later.

Below: Two of the Short 827 Type seaplanes after conversion to landplanes at Basra.

promptly manoeuvred into position to attack. The observer released four 16-lb bombs in a direct line with the submarine at a distance of about 100 feet apart. The first two appeared to drop short but the remaining two fell close to the vessel. It was difficult to determine whether the submarine had actually been hit. A dark smoke was observed in the place where the submarine submerged on both occasions.[71]

On Thursday 24 May 1917, Sub-Lt H.M. Morris was ordered, with AM2 G.O. Wright (W/T) as observer, to go in search of Flt Sub-Lt Maxton who had failed to return in his Sopwith Baby N1065 and must have been forced down onto the sea. The Station did not know that this seaplane had been picked up and taken in tow by *Hopper 15* from the Isle of Grain and was safely in harbour before No. 3072 began the search for her over the North Sea.

Morris was forced down after about 30 minutes when *my engine suddenly stopped. I was forced to land. The sea was choppy and the wind rising, so my observer sent off his pigeon while I kept the machine head to wind.*

At about 2.30 my starboard lower plane was carried away causing us to swing broadside on to the sea: we climbed out on to the other plane so as to balance things, but the machine gradually began to sink tail first. As she sank we climbed out on to the floats and sat on them, till I was washed off, but managed to catch the tail under water and climbed on again. Here we sat till the machine sat propeller in air and finally turned right over.

The wind was blowing a gale and the sea was very rough. The following morning the wind dropped considerably. After sunset six British seaplanes flew over and they answered the waving crew by firing a green light. Nothing further happened until Tuesday at about 1400 when a Curtiss H-12 saw them and alighted although the sea was getting rough again. As it passed the pair they clung onto its wires and climbed in. The Curtiss tried to take off but it was too rough and only broke the tailplane. The flying boat then taxied about 25 miles until they were sighted by the *Orient* and picked up.

To the pilot's report was added the following: - *Not often have words so bald been used to cover a tale of comparable endurance. All that the pilot and his observer had to eat during the five days and nights they were clinging to the float of the wrecked seaplane were a few malted milk tablets. One of the pigeons which were released was never seen again, but the other fell exhausted on the deck of a mine-sweeping trawler and, after being succoured, was released once more and eventually found its way to its loft at Westgate.*

This pigeon did not return to its loft until the third day and there was no record that any action had been taken on receipt of the message. The seaplanes that flew over the castaways on the 25th and fired a green light were not British. German aeroplanes had attacked Folkstone that afternoon and many aeroplanes went up in pursuit. It was considered unlikely that the seaplanes were German. *There are recorded instances when German pilots risked their lives to rescue British airmen in peril on the North Sea, but none when they callously left them to their fate. It is not impossible that the seaplanes were a phantom Flight, created from the stuff of the hopes and sufferings of the two airmen.*[72]

The flying boat was one of two that had left Felixstowe on an anti-submarine patrol when they were separated due to heavy fog, one boat turning back. The other, with crew of Flt Sub-Lt J.L. Gordon and Flt Sub-Lt G.R. Hodgson, Leading Mechanic S.F. Anderson and W/T Operator B.W. Millichamp. They saw something on the water through a break in the fog and investigated and observed it to be two people on an upturned Short float. They climbed to 1,000 feet and sent a message to base that they were going to alight to pick up survivors. They dropped three of their bombs, the fourth hung up. They landed close by and on their second attempt managed to get the crew aboard by taxying directly up to them and catching them on the drift wires. They made two attempts to take off and on the second smashed the tail plane and starboard wing float. While taxying the tail was leaking badly and the boat was shipping considerable water the bows, so much so that the boat made a distress signal and was taken in tow by the *Orient of Leith* that was bound for Yarmouth.[73]

In 1917 a small number of 827 Type seaplanes ended up in the Mediterranean being based at Otranto.

Seaplane Training

The F.B.A. flying boats was used for training on Lake Windermere. While considered adequate for initial training with their side-by-side seating, they were like no seaplane that they would serve on, so Commodore Paine ordered three advanced trainers – Short 827 seaplanes. No. 3328 (Brush-built) and No. 8560 (Fairey-built) were assembled by workmen from the two companies October 1916 at Hill of Oaks. The two groups competed to see who would get their machine completed first, the Brush workmen 'winning'. On 25 October, another Brush-built arrived – No. 3323, delivered to the Windermere railway station from Grain. No. 3328 gave the first instructional flight by the 827 Type on 26 November. No. 8560 had failed its acceptance test due to a faulty magneto.

RNAS Windermere ceased active operations in January 1917. No. 8560, that had overcome its magneto problems, was the last serviceable machine and made the last recorded flight on 13 January, and was sent by rail to the RFC School of Aerial Gunnery at Loch Doon, becoming A9920. Of the other 827 Types No. 3328 was scrapped at Hill of Oaks after a crash in early December. No. 3323, that had been assembled in January but was waiting for its dual controls, was taken to the railway station for shipment to Calshot.[74]

A Meeting was held on 14 January 1918, with regard to Seaplane Training. The system of training at Lee-on-Solent and Calshot was described by Wing Cdr Bigsworth:

Pupils were first put on Short 827 type then passed to F.B.A. Boat and then to N.T.2.B type: they were then passed out and did a Gunnery Course on Short Type Seaplane.

Left: Great Yarmouth Short 827 No. 3105 has the serial repeated on the fin. The engine panels appear to have been painted to match the doped fabric. No. 3105 was at Yarmouth by 27 March 1916. It participated in an anti-Zeppelin patrol on 19 June. Deleted 31 October 1916.

Right: Working on the 150-hp Sunbeam engine of No. 3105 at Great Yarmouth. Note the pipes from the side radiator.

Left: View of Short 827 and Sopwith Schneider seaplanes at Great Yarmouth, 1916. Nos. 8222 and 8223 identified thanks to the large serial on the fin. Both were erected at Yarmouth in February 1916.

Right: Short 827 No. 8223 with Short 184 No. 8385, and other unidentified Short and Sopwith Baby seaplanes. Both Shorts served at Yarmouth.

Left: Short No. 8233 with three-bay equal span wings. Possibly taken at Grain. Note the application of a Protective Covering (PC) doping scheme with the engine panels painted to match.

Right: Short No. 8561 in PC dope with grey metal panels. Note the three-bay equal wings and the wing tip floats.

Left: No. 8231 showing how the tail had to be propped up with the two-wheel trolley arrangement. Delivered to HMS *Campania* by March 1917. At Lee-on-Solent in April 1918, it was probably used for training. Deleted the W/E 27 June 1918.

Right: Short No. 3331 showing the large trolley used at Westgate for launch and recovery. The seaplane required the rear fuselage to be held up until the machine was in the water. The ground crew are holding the rods that the wings rest against when folded. No. 3331 attacked a German submarine near the Kentish Knock on 23 May 1917. At the training school at Lee-on-Solent by November 1917. Deleted W/E 12 September 1918.

Left: Short No. 3331 after landing on the grass covered field without damage, just embarrassment. Note how the exhaust from the 150-hp Sunbeam engine ejected at the bottom of the fuselage.

Right: Short No. 3326 taxying at Calshot on 21 January 1918. Delivered to Calshot in July 1917, it appears to have spent the reminder of its service life at this station until deleted in March 1918.

Below: No. 8556 has the same type of serial application - possibly a Calshot hallmark? The company logo appears to be carried on the central rudder stripe. No. 8556 came to an unhappy end on 20 July 1917, when it hit the mast of a transport ship while alighting in Southampton Water. It dived onto the deck of the ship killing the crew of Flt Sub-Lt Ian Neil Carmichael and Flt Sub-Lt William Ross Wallace. (Sturtivant & Page record the observer as A Wallace.)

Above: Short No. 8555 shows another serial number application. This machine also served at Calshot.

Left: Short No. 8229 was fitted with dual control for training. It was at Lee-on-Solent by September 1917, with A Flight. Note the louvres in the engine panels.

Left & Above: Short View of the training machines on the beach at Lee-on-Solent – Short 827 and 184 seaplanes, and F.B.A. flying boats. Note the construction work in progress.

Right: A Short 827 at Lee-on-Solent after some rough handling.

Above: Fairey Aviation Co Ltd constructed the batch Nos. 8550 – 8561. Of these No. 8560 became A9920 after transfer to the RFC in early 1917 for work at the School of Serial Gunnery, Loch Doon., the only Short 827 Type to serve in the RFC. It is seen on Lake Windermere where it was sent for testing by Sydney Pickles. Note the fixed wingtip floats.

Right: Lewis gun mounted on a Short 827. Where and when unknown. It appears to only be able to be fired by the pilot straight ahead over the top plane.

Left: A Short 827 by a shingle beach.

Right: A Short 827 on the point of lifting from the water.

Left: No. 8634 demonstrating another type of tail support. A lot of the mundane but necessary items for the safe storage and movement of seaplanes on RNAS stations appear to be locally made items. No. 8634 was with the Nore War Flight at Felixstowe in August 1916. At the Killingholme seaplane school by July 1917, it was deleted in the W/E 22 February 1918.

On passing out of the Gunnery School, they proceeded to patrol Stations for flying twin Float type, and when 20 hours patrolling had been completed, they were eligible for Large America Training.

If selected for Large America training a course in these boats was taken followed by a gunnery course in these machines. This system was generally approved by the Meeting. The following points were emphasised: -

(a). Small Americas were not required, as they are harder to fly than Large Americas.

(b). Short 260 Sunbeam were unsuitable for instructional purposes, as they were too high powered and the seaplane would not stand the rough treatment given to it by the average pupil.

(c). A small instructional boat is required, but the F.B.A. is a very bad specimen of the type.

An *Instructional Float Type Tractor Seaplane. To be roughly similar to Short 827 type with 200 Hispano-Suiza or "Arab" engine* was required.[75]

A float type school machine was considered an absolute urgent necessity. While it was considered that the float type seaplane was disappearing, they were still required to meet the output of pilots that had to be trained. *The present type of war machine is quite unsuitable for elementary training purposes, and whatever the future policy is that type school machines will always be useful.* 30 float seaplanes were considered necessary for Lee-on-Solent and 12 for Calshot.[76]

At the 33rd Meeting of the Admiralty Air Department Progress Committee on 22 January 1918, the Seaplane Training tractor float seaplane was discussed in the light of the above Meeting, and it was suggested that the old 150 Short be revived, however, it was noted that there were objections to this proposal as a great many parts were not standardised. Capt A.V. Vyvyan suggested that the 827 Short was the most satisfactory machine available and a large number would not be required. *The engine for this type was discussed, and the 170 Amazon Sunbeam, or the 190 Renault appeared the best proposition, there being about 100 of either type available.* The matter was to be looked into.[77]

To the question 'Is an instructional floatplane required?' an unnamed RNAS Station replied that *observing that Short 827 and 166 Types are rapidly dying,* the answer was 'Yes', to an *827 Short type with 200-hp H.S. or Arab engine.*[78] The *Sage Converted Float seaplane* was put forward as a machine that would be able to meet the requirements for a school machine. It was proposed to build two quickly as the parts were in store at the firm's works. These would be the two Sage 4 seaplanes Nos. N116 and N117. If they were successful, the production models would be available about the end of July 1918. *It is considered that this time will not be shortened by the placing of production orders for Short 827 Type, taking into consideration the lack of drawings and the fact that the engine installation would require re-designing.*[79]

Only 108 Short 827 seaplanes were constructed. According to the Admiralty Disposition of Aircraft on Naval Duties dated 28 November 1918, the following Short 827 floatplanes were still in service: -

- At Bogton A9920 (stored).[80]
- At Lee-on-Solent 8237 and 8261.
- At Port Said 8649.
- All were fitted with 150-hp Sunbeam Crusader engines.

No. 8560 was obtained for the School of Aerial Gunnery at Loch Doon, and given the RFC serial A9920, the only Short 827 to be operated by the RFC.

Disposition of Short Seaplanes on Seaplane Carriers 31 May 1915 Short 827 in Belgian Service

While the Belgian forces were confined to protecting their homeland in WWI, at that time the Belgian Congo was partially separated from German East-Africa by Lake Tanganyika. The Germans had two steamships, the *Graf von Götzen* and the *Hedwig von Wisman* that were to be fitted with guns salvaged from the *Koenigsberg,* and some smaller vessel on the Lake. The Germans conducted a campaign against the Belgians with the support of the native tribes. Belgian did the same and finally decided to invade German East-Africa. In April 1916, the Belgians launched a successful attack from north of the Lake. However, it was realized that as long as they did not control the Lake it could not be used for transport, which meant that they could not support their own troops. The two gunboats were shelling the Allied outposts with impunity. Their smaller craft were also making nuisance raids on the Lake. *We are constrained by the Graf von Götzen with its superior calibre weapons.* The destruction of this vessel was the *underlying theme in all our offensive operations.*[81]

General Tombeur, one of the pioneers of the Congo, had, a small force armed with obsolete equipment, to cover an immense area. He knew that he had to gain command of Lake Tanganyika and to achieve this he needed aircraft or, ideally, a submarine.

HMS *Campania*

Serial	Type	Engine	Notes
823	Short 827 Type	150 Sunbeam	Ready.
824	Short 827 Type	150 Sunbeam	Ready.
826	Short 827 Type	150 Sunbeam	Ready.
925	Sopwith 807 Type	100 Gnome	Ready.
926	Sopwith 807 Type	100 Gnome	Ready.
1559	Sopwith Schneider	100 Gnome	Allocated.
3707	Sopwith	100 Gnome	Allocated.
836	Wight 840 Type	225 Sunbeam	Erecting.
837	Wight 840 Type	225 Sunbeam	Erecting.
838	Wight 840 Type	225 Sunbeam	Ready. In transit.

HMS *Engadine*

Serial	Type	Engine	Notes
819	Short 830 Type	135 C Unné	Ready
852	Sopwith 860 Type	225 Sunbeam	Ready
858	Sopwith 860 Type	225 Sunbeam	Ready
1336	Short 830 Type	135 C Unné	Ready
1443	Sopwith Schneider	100 Gnome	Ready

HMS *Rivera*

Serial	Type	Engine	Notes
1335	Short 830 Type	135 C Unné	Ready.
1337	Short 830 Type	135 C Unné	Allocated.
1338	Short 830 Type	135 C Unné	Allocated.
3067	Short 827 Type	150 Sunbeam	Allocated.

HMS *Ark Royal*

Serial	Type	Engine	Notes
136	Short 135 Type	200 C Unné	Ready
161	Short 166 Type	200 C Unné	Ready
172	Wight Improved Navyplane	200 C Unné	Ready
173	Wight Improved Navyplane	200 C Unné	Not ready.
176	Wight Improved Navyplane	200 C Unné	Ready
807	Sopwith 807 Type	100 Gnome	Ready
857	Sopwith 860 Type	225 Sunbeam	Ready
860	Sopwith 860 Type	225 Sunbeam	Ready
922	Sopwith 807 Type	100 Gnome	Ready
1201	Sopwith Tabloid	80 Gnome	Ready
1202	Sopwith Tabloid	80 Gnome	Ready
1203	Sopwith Tabloid	80 Gnome	Ready
1204	Sopwith Tabloid	80 Gnome	Ready
1437	Sopwith Schneider	100 Gnome	Ready
1438	Sopwith Schneider	100 Gnome	Ready
162	Short 166 Type	200 C Unné	En route.
175	Wight Improved Navyplane	200 C Unné	En route.

HMS *Ben my Chree*

Serial	Type	Engine	Notes
184	Short 184 Type	225 Sunbeam	Ready.
185	Short 184 Type	225 Sunbeam	Allocated
820	Short 830 Type	135 C Unné	Ready.
821	Short 830 Type	135 C Unné	Ready.
841	Short 184 Type	225 Sunbeam	Ready.
842	Short 184 Type	225 Sunbeam	Allocated
1445	Sopwith Schneider	100 Gnome	Ready
1560	Sopwith Schneider	100 Gnome	Ready.
1561	Sopwith Schneider	100 Gnome	Ready.

Note: Machines also operated from HMS *Minerva* and *Doris* as required. HMS *Empress* was refitting, and HMS *Vindex* was being fitted out as a seaplane carrier.
Source: TNA ADM1/8021.

This table shows the few Short 827 Types (four) compared with the Short 830 Type (seven) on Seaplane Carriers at this time. There were 15 miscellaneous Shorts out of the 25 two-seat seaplanes.

Tombeur finally obtained permission to establish an aviation unit for the colony and Lt Albert de Bueger was ordered to form such a unit. De Bueger had served in the colony and knew it well, also he was the pilot of the *Aviation Militaire Belge's* sole seaplane. He chose the Short 827 floatplane and four were acquired from the British Admiralty – Nos. 3093, 3094, 3095 from the Sunbeam Motor Car Co, and No. 8219 a Parnell built machine.

According to Pieters the aircraft were shipped from London on the Belgian steamship *Anversville*. However, Sturtivant and Page record that they were delivered to Grain and accepted in November and December 1915, then shipped to No. 8 Squadron at Zanzibar, East Africa. Once in Africa they were transferred to the Belgians.[82]

The *Anversville*, left the UK on 7 January 1917. The ship carried two spare engines, 1,000 lbs of bombs, four machine guns and 30,000 rounds of ammunition as well as 70,000 litres of aviation fuel. A fire on board meant the loss of about a third of the fuel. The ship arrived at Matadi on 4 February. The aircraft and supplies now had to be carried across to M'Toa on LakeTongwe where a seaplane base would be established. Lake Tanganyka had proved too open and windy for the seaplanes to operate from its waters and the quite waters of Lake Tongwe were chosen instead. The crated aircraft and supplies had to be manhandled overland to the base where they were erected and tested.

The first aircraft was not ready for flight until 14 May. *Souis Lt* Aimé Behaeghe performed taxi runs until he was sure of the machine then, after a long take-off run, took to the air. The unit then began to train for its primary task, the

Above: HMS *Ben-my-Chree* hangar area with a Short 184 seaplane on the crane.

Right: Short 827 Type from the Sunbeam batch Nos. 8630 to 8649, No. 8637 has the arched spreader bars associated with torpedo seaplanes but with a bomb carrier mounted under the fuselage. This seaplane served at Yarmouth from June 1916, but was deleted on 17 October 1916, after crashing to a wreck while alighting on 5 October.

Left: A 827 Type slowly taxying on the water. Note the missing engine panels.

Left: An 827 Type taxying. As speed built up, as shown by the wake, the ailerons moved up to the neutral position.

Right: A Short 827 Type at Eastchurch. Note the metal panels, floats and struts are all the same colour. The rear engine panels have louvres.

Belows: A Fairey built Short 827 Type.

Above: HMS *Ark Royal* would be closely associated with Short floatplanes throughout the war.

Above: This photograph shows details of the forward fuselage of No. 3063, note the auxiliary oil storage tank on the fuselage side, the flat engine panels without cooling louvres, exhausts for the 150-hp Sunbeam engine, and the metal footsteps attached to the undercarriage cross bracing wires. (RAFM P008886.)

Above: In this view the engine panels, undercarriage struts and floats appear to be white. The double struts at the side of the fuselage associated with the wing fold are well shown. On the original print, louvres may be discerned on the intermediate engine side panel. (E.M. King Collection, Puke-Ariki Museum).

sinking of the German steamship *Graf von Götzen*. On 2 June the first accident occurred, *Sous Lts* Ort and Castiau coming to grief. Their floatplane had to be recovered and repaired. A bombing flight on the 11th saw the seaplane not return that evening. It had forced alighted and been towed into Albertville by the Belgian gunboat *le Vengeur*. The same crew, Sous Lt Behaeghe and Lt Colignon, succeeded on the 12th in bombing the *Graf von Götzen* and placing it out of action for the rest of the campaign.[83]

The two seaplanes were joined by the third on 26 June and the trio were kept busy in carrying out reconnaissance, photographic and bombing flights. On 26 July, seaplane Nr. 3 crashed on the return from a raid. The fourth machine was erected and operating by the 30th. At least one Short was returned to Europe and operated from Calais, possibly No. 8219.[84]

Endnotes

Right: E.M. King with his Kodak camera. The 827 Type in the background appears to be the same aircraft as above. (E.M. King Collection, Puke-Ariki Museum).

This Page: A series of photographs in a trilingual paper shows the Short 827 seaplanes in Belgian service at Lake Tanganyika. No. 3094 can be recognized in the original, still in its RNAS markings.

Left & Belows: A series of photographs in a trilingual paper shows the Short 827 seaplanes in Belgian service at Lake Tanganyika. No. 3094 can be recognized in the original, still in its RNAS markings.

Left: The hut like hangars constructed for the Belgian Short 827 seaplanes. Note that the aircraft in the background has its wings fitted but folded.

Right: Two of the Belgian Short 827 seaplanes showing a mixture of markings. The far one has Belgian cockades inboard of the ailerons with the British ones still applied. The machine nearest the camera retains British markings but the fuselage cockade in now in Belgian colours.

Left: Launching a Belgian Short 827 on Lake Tungwe. The rudder stripes appear to have been altered to Belgian colours.

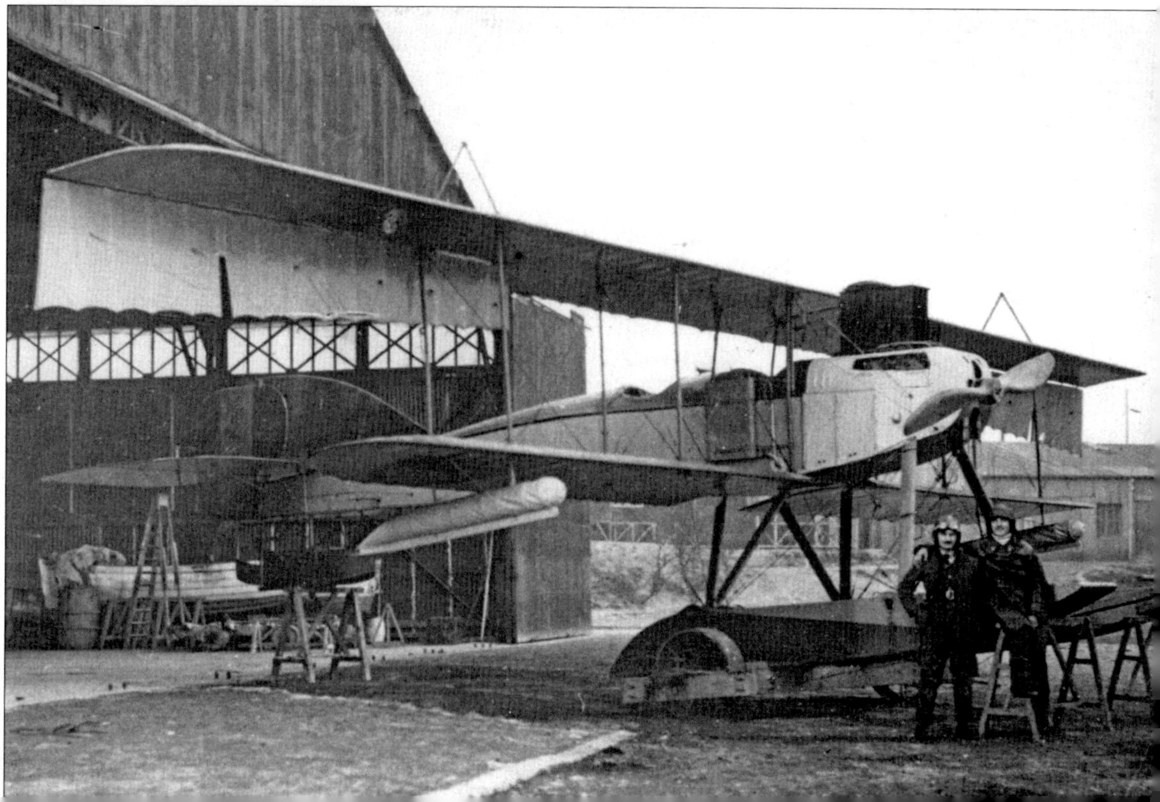

Right: This Short was photographed at Calais in February 1918. This is a survivor of the Short 827 seaplanes from the Belgian Congo. Truly a remarkable longevity for such a machine.

1. Delivery dates from Sturtivant, R & Page, G. *Royal Navy Aircraft Serials and Units 1911-1919,* Air Britain, UK, 1992.
2. For example: Bruce, J.M. *British Aeroplanes 1914-1918.* (Putnam, UK, 1957) P.483.
3. Log Book on Nos. 3097 & 3100. TNA AIR1/2072/204/412/3097 & AIR1/2072/204/412/3100.
4. When Richard Fairey left Shorts to start his own company, Horace Short promised him a sub-contract for twelve 827 Type seaplanes. What Horace did not know, according to Harold Penrose, was that Fairey had taken with him not only drawings of the 827 Type that was then being test flown by Roland Kem, but a folio of other projects including a large single-engine bomber. Penrose, H. *British Aviation – The Great War and Armistice*, Putnam, UK, 1967. P.57.
5. Log Book on Nos. 3097 & 3100. TNA AIR1/2072/204/412/3097 & AIR1/2072/204/412/3100.
6. Flt S/Lt John James de la Tour Fox and AMII Robert Edward Gorman from Portland were lost in Short 184 No. 8367 in the Channel on 9 March 1917. Fox was 24 and Gorman just 19 years of age.
7. Report titled 'Portland', received from Capt C.H. Keith, RAF (S.O.2.Air) No. 10 (Naval) Group, Warsash. 07.06.1919. Apparently, it was for a history of Portland NAS. TNA AIR1/718/28/3.
8. 'Report on proceedings. Seaplane 8379 assisting H.M.S. Q.7. in destroying enemy submarine U.19.' TNA AIR1/721/59/3/6.
9. Report titled 'Portland', received from Capt C.H. Keith, RAF (S.O.2.Air) No. 10 (Naval) Group, Warsash. 7.06.1919. Apparently, it was for a history of Portland NAS. TNA AIR1/718/28/3.
10. Snowie, J.A. *Collishaw & Company.* Nieuport Publishing Ltd, Canada, 2010. P.61.
11. Grain – 'Summary of Operations & Reconnaissance. Period Ending April 28th 1916.' TNA AIR1/436/15/278/1.
12. TNA AIR1/436/15/285/1.
13. *LM* possibly Leading Mechanic.
14. RNAS Operations Reports Parts 1 to 36. The Naval & Military Press Ltd, UK, 2020. P.225. Sturtivant & Page give the spelling as McNicoll.
15. See Moore, W.G. *Early Bird,* Putnam, UK, 1963, for Moore's full story.
16. Lincoln's Diary entry for 23 March 1916. Lincoln's Diary via P Dye.
17. Moore, W.G. *Op Cit.* P.30.
18. Lincoln Diary. *Op Cit.*
19. Lincoln Diary. Op Cit. Entry for 4 May 1916.
20. Lincoln Diary. Op Cit. Entry 10 April 1916.
21. Robert Dowty Pershore. Worcestershire, England. LM RNAS 8744. From title page of diary. FAA Museum.
22. HMS Vengence was a *Canopus*-class pre-dreadnought, not a cruiser.
23. Lincoln Diary. *Op Cit.* Entry for 13 April 1916.
24. Lincoln's Diary. *Op Cit.* Entry for 24 Apr 16. In the Cape Station's report for the 2 to 26 June 1917, special mention was made of No. 8253. Overhauling completed and Flt Cmdr Maclean reached 6,400 ft in 1:35 hrs and on a 2nd patrol reached 7,400 ft in 2:05 hrs. A seaplane record for East Africa. RNAS Operations Reports, P.31. *Op Cit.*
25. Gallehawk, A.N. *Op Cit.*
26. Report of Operations No.16. Extract from Report of No. 1 Wing. *RNAS Operational Reports No2. 1 – 29, 26th Nov 1915 to 16th March 1917.* P.67. TNA AIR1/2314/223/12.
27. No. 3097 was still carrying out hostile craft patrols in August 1917. There is a gap in the Log Book until March 1918 when it undertook another such patrol. The machine was *deleted from active list 15th March 1918.* An exceptional record for a 1915 seaplane operating in tropical conditions.
28. 'Report of Operations of No. 8 Squadron at Zanzibar.' Copy in FAA Museum 1984/127/0043.
29. Report of Operations No.19. Extract from Report of No. 1 Wing. *RNAS Operational Reports No2. 1 – 29, 26th Nov 1915 to 16th March 1917.* P.91. TNA AIR1/2314/223/12.
30. Gallehawk, A.N. *Op Cit.*
31. Report of Operations No. 21. Seaplanes (Zanzibar). P.113. *RNAS "Operations Reports."* Reprint by Naval and Military Press, UK, 2019.
32. In his recollections Gallehawk records that Bowhill arrived on 8 August and the squadron embarked for England on 12 August, reaching Plymouth on 9 September 1916. Gallehawk, A.N. *Op Cit.*
33. Translation of letter from the Khalifah of Zanzibar, dated 11.08.1916. FAA Museum Album 101. B0416/0001.
34. The SS *Mombasa* was sunk on 20 Oct 16 in the Mediterranean off Cape Corbelin, Algeria by the U.39, en-route from London to Zanzibar. Note to Diary by P Dye.
35. Lincoln Diary. *Op Cit.* Entry for Sep-Dec 1916.
36. Report by HMS *Himalaya* for 11.11.1916. Copy in R Moon's file. FAA Museum. There are many such reports in this file and this is typical, most flight starting early. 8642 was completely wrecked on 8 January 1917, while flying from Zanzibar to Dar es Salaam.
37. RNAS Operations Report No.29. *RNAS Operations Reports*, Naval & Military Press, UK, 2019. P.241.
38. The 'RNAS Operations Report for 1st to 16th April 1917.' (Admiralty War Staff, Operational Division, 17 April 1917.) lists No. 8254 with No. 8 Squadron at the time of its demise.
39. 'Story of Flight Commander Moon.' FAA Museum File 1984-127.

Short 827 Serial Allocation

Serial	No.	Contract No.	Manufacturer	Notes
822 – 827	6	CP57105/14	Short Bros	
3063 – 3072	10	CP43399/15	Short Bros	3063 to 3065 were with Nore Flight on 24.04.1916.
3093 v 3112	20	CP49441/15 & CP70484/15	Sunbeam Motor Car Co.	3106 & 3107 were with Nore Flight on 24.04.1916.
3321 – 3332	12	CP32600/15	Brush Electrical Engineering Co.	
8218 – 8229	12	CP48104/15	Parnall & Sons	
8230 – 8237	8	CP32600/15	Brush Electrical Co.	Nos. 8231, 8233 & 8237 three-bay version.
8250 – 8257	8	CP76208/15	Parnall & Sons	Additional eight 150-hp seaplanes.
8550 – 8561	12	CP57387/15	Fairey Aviation	8560 to RFC.
8630 – 8649	20	CP78661/15	Sunbeam Motors	

Specifications Short 827

Source	1	2	3	4
Span upper	54 ft 0 in	53 ft 11 in	53 ft 11 in	53 ft 11 in
Span lower	36 ft 0 in			40 ft
Length	38 ft 5 in	35 ft 3 n	35 ft 3 in	35 ft 3 in
Height	13 ft 3 in	13 ft 6 in		13 ft 6 in
Chord				5 ft
Gap				5 ft 6 in
Stagger				0°
Dihedral	Nil			0°
Airscrew Diameter	10 ft 4 in			
Airscrew Pitch	80 ins (mean)			
Wing Area, ft^2	448		506	
Empty Weight, lb.	2300 (approx.)[1]		2,700	3,400
Military load, lb.				
Crew, lb.				360
Fuel & oil, lb.				
Main floats, lb.	215 each			
Loaded Weight, lb.			3,400	
Main fuel, gal.	36			
Auxiliary fuel, gal.	10			
Oil, gal.	6			
Speed max			62 mph	61 mph
Endurance			3.5 hr	
Engine	Sunbeam Coatalen 150-hp			150-hp Sunbeam

Notes: 1. Without fuel or crew.

Source: 1. Log Book on Nos. 3097 & 3100. TNA AIR1/2072/204/412/3097 & AIR1/2072/204/412/3100.
2. Types of British Seaplanes, Flying Boats and Ship's Aeroplanes. TNA AIR10/96.
3. C.H. Barnes data.
4. J.M. Bruce data.

40. 'Flight over the Rufiji Delta in search of missing seaplane No. 8254'. Report by Flt Lt Sub-Lt E.E. Deans. 06.01.1917. TNA AIR1/2399/283/2.
41. Sturtivant and Page (*Op Cit*) show the date as the 8 January 1917. Lincoln's diary is considered correct. The remains of No. 8642 were brought to the Station on the 8th.
42. Court of Enquiry, Seaplane Accident, 6th January 1917, Rufiji Delta, 11.02.1917. TNA AIR1/2399/283/2.
43. RNAS Operations Report No. 31. 1st to 16th April 1917. Copy in FAA Museum File 1984-127 on R Moon.
44. Report by Acting Flag Capt Cyril Goolen, HMS *Talbot*, Dar es Salaam17.04.1917. TNA AIR1/2399/283/2.
45. Note to copy of report. June 1918. TNA AIR1/2399/283/2.
46. RNAS Report of Operations for period 17 to 31 July 1917, Cape Station, No. 8 Squadron.
47. RNAS Report of Operations for period 2 to 26 June 1917. Cape Station, No. 8 Squadron.
48. RNAS Report of Operations for period 1 to 16 August 1917, Cape Station, No. 8 Squadron.
49. Bruce, J.M. *British Aeroplanes 1914-1918,* Putnam, UK, 1957. P.485. Sturtivant & Page *(Op Cit)* list Nos. 3093 – 3095 and 8219 as transferred from No. 8 Squadron to the Belgians.
50. Lincoln Diary. *Op Cit*. Entry for Mar-Sept 1916.
51. 'Extract from an Official Report made by British Officer to the Director of Air Service.' No Date. This report may have been made by Wing Cdr Bowhill (later Air Chief Marshall Sir Fredrick Bowhill, GBE, KCB, CMG, DSO & Bar) circa February 1918. Copy in NARA RG45 Entry 520 Box 138. English Air Operations-Air Operations other than Mediterranean, Air War North Sea.
52. For a modern view see Lax, M; O'Connor, M & Vann, R. *Wings over Mesopotamia, Air War in Iraq 1914-1918,* CCI & ASWWIAH, UK, 2017. This book is profusely illustrated with photographs of the aircraft of both sides in the conflict.
53. Townshend went into relative luxury to sit out the war while his men suffered greatly. Townshend was unable to understand why his friends and Army comrades turned against him when his behaviour became known.
54. Sueter, M.F. *Airmen or Noahs*, Sir Isaac Pitman & Sons, Ltd, UK. 1928. P.59.
55. 'War Experiences' Flt Lt A.C. Bishop. AH.228/11/35. (2nd Course).
56. 'War Experiences' Flt Lt A.C. Bishop. *Op Cit.* For an overview of the climatic conditions in Mesopotamian See Jones, H.A. *The War in the Air, Vol. V.* P.250.
57. Report by T Gordon, Squadron Commander. 'Report on Naval Air Service Operations in the Persian Gulf Sept – Oct 1915.' TNA AIR 1/665/17/122/711.
58. Sturtivant, R & Page, G. *Op Cit*. P.46.
59. Jones, H.A. Vol. V. *Op Cit.* P.258.
60. 'General Work Done.' Probably May 1916. FAA Museum File: 2010/112/0010.
61. Sueter, M.F. *Airmen or Noahs*, Sir Isaac Pitman & Sons, Ltd, UK. 1928. P.60.
62. The monoplane was probably a Fokker or Pfalz E type fighter fitted with a synchronised machine gun. TNA AIR1/149/15/103.
63. Henshaw, T. *The Sky Their Battlefield II,* Fetubi Books, UK, 2014. P.295.
64. Cowton, A.E. *With the First in the Field*, Mrs A.E. Cowton, UK, 1963. P.158. Cowton served from 1905 to 1929 in the Royal Marines Light Infantry, RNAS on Kite Balloons, and in the RAF in No. 27(B) Squadron on the North-West Frontier. His autobiography was privately published by his wife.
65. Bruce, J.M. *British Aeroplanes 1914-1918,* Putnam, UK, 1957. P.485. See also Jones, H.A. Vol.V. *Op Cit*. P.268.
66. 'Extract from an Official Report made by a British Officer to the Director of Air Services.' The report was signed by a Wing Cdr, RN (CO, RNAS) and was sent to the Director of Air Services, RN, Air Sept. No date. NARA RG45 Entry 520 Box 152. This report may have been made by Wing Cdr Bowhill (later Air Chief Marshall Sir Fredrick Bowhill, GBE, KCB, CMG, DSO & Bar) circa February 1918.
67. TNA AIR1/150/15/107.
68. Cole, C & Chessman, E.F. *The Air Defence of Great Britain 1914-1918,* Putnam, UK, 1984. P.127.
69. Jones, H.A. *The War in the Air*, Vol. 4 P. 70.
70. *RNAS Operations Reports,* Reprint, Naval & Military Press, UK, 2019.
71. *Ibid*.
72. Air Episode in the North Sea, May, 1917.' TNA AIR1/630/17/122/25. Westgate became the first Naval Air Station to establish a pigeon loft, this occurring in the summer of 1916. See Williams, G. 'Wings over Westgate. *Cross & Cockade International.* Vol. 18 No. 3, 19187. P.126.
73. *Ibid.*
74. Connon, P. *An Aeronautical History of the Cumbria, Dumfries and Galloway Region Part 2: 1915 to 1930.* St Patrick's Press, UK. 1984. P.91-99. Connon also gives a history of the setting up of the RFC School of Aerial Gunnery.
75. Minute No. 14365. Seaplane Training. To ADAS & ICAT from C Courtney(?), N.1. 15 January 1918. AIR1/664/17/122/695.
76. Note to Minute No. 14365. *Ibid.*
77. TNA AIR1/270/15/266/113.
78. Seaplane Training. Undated, unsigned memorandum. About January 1918. TNA AIR1/664/17/122/695.
79. Minute from Controller, Technical Dept to DAS and N.1. 23.01.1918. TNA AIR1/664/17/122/695.
80. Bogton was the aerodrome near Dalmellington that was used by the landplanes of the Loch Doon school.

81. Note by Col Moulaert, 13 March 1916. Quoted *in CCI Vol. 39/1*, 2008, P.10.
82. Pieters, W. *The Belgium Air Service in the First World War*, Aeronaut Books, USA, 2010. Sturtivant, R & Page, G. *Royal Navy Aircraft Serials and Units 1911-1919.* Air Britain, UK, 1992. The following account is based on Pieters research.
83. In his article 'The Battle of the Lake,' (*RAF Flying Review*, Vol. XVI, No.2. P.35), Victor Houart records that Coligon's attack was not carried out as the machine had to alight with engine trouble and was towed back to base. He credits Orta with the successful attack. Work of repairing the *Graf von Götzen* was apparently abandoned.
84. It is not known if the Belgium serials were given in the same sequence as the British serials.

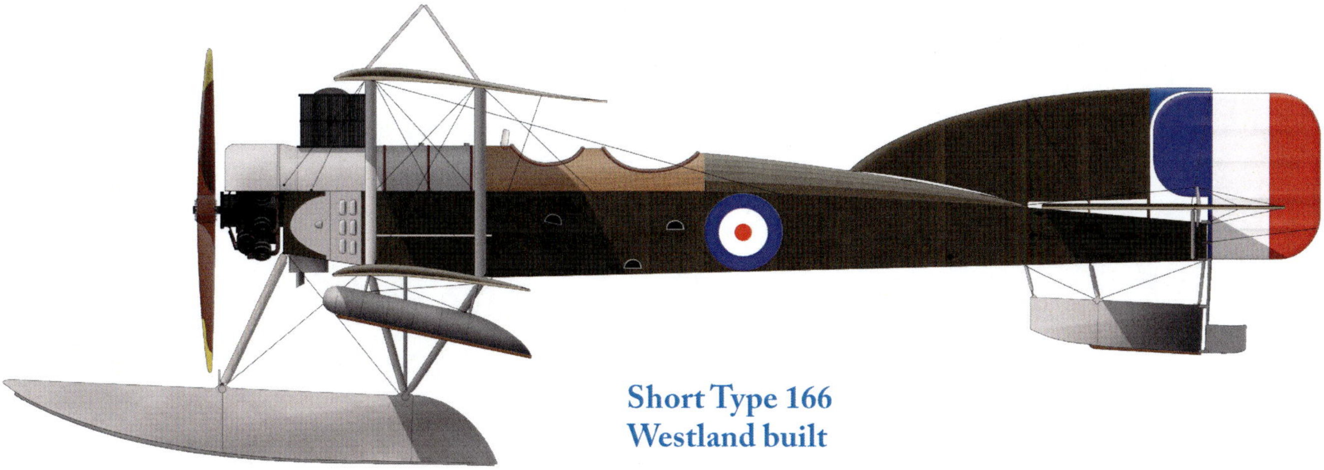

Short Type 166
Westland built

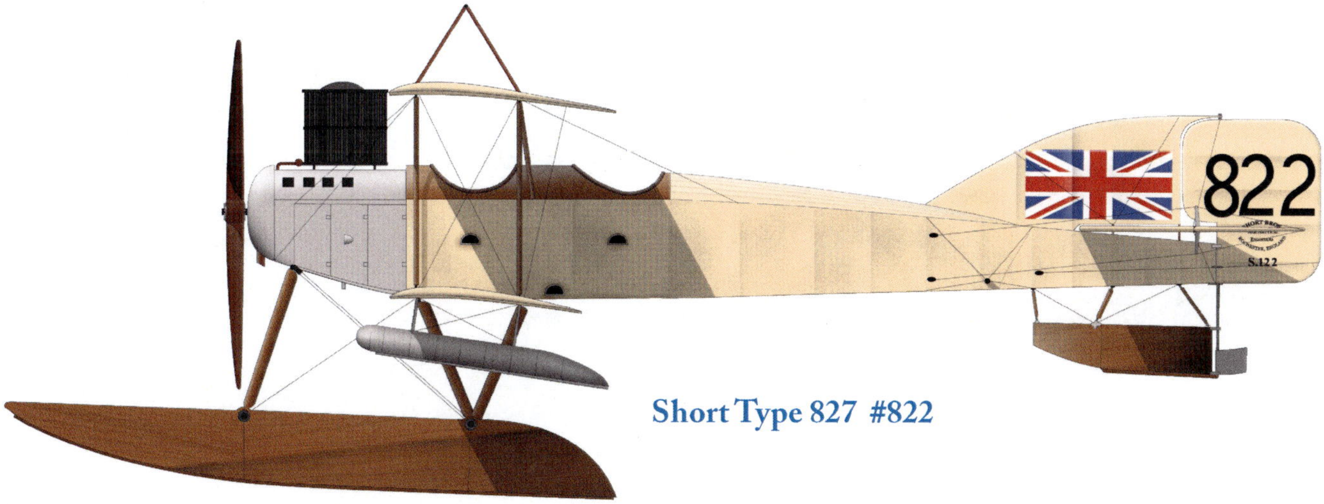

Short Type 827 #822

Short Type 830 #9782
Lee-on-Solent NAS
Summer 1917

Short 166 Type Number 162
HMS *Ark Royal*

Short 166 Type Number 166
HMS *Ark Royal*

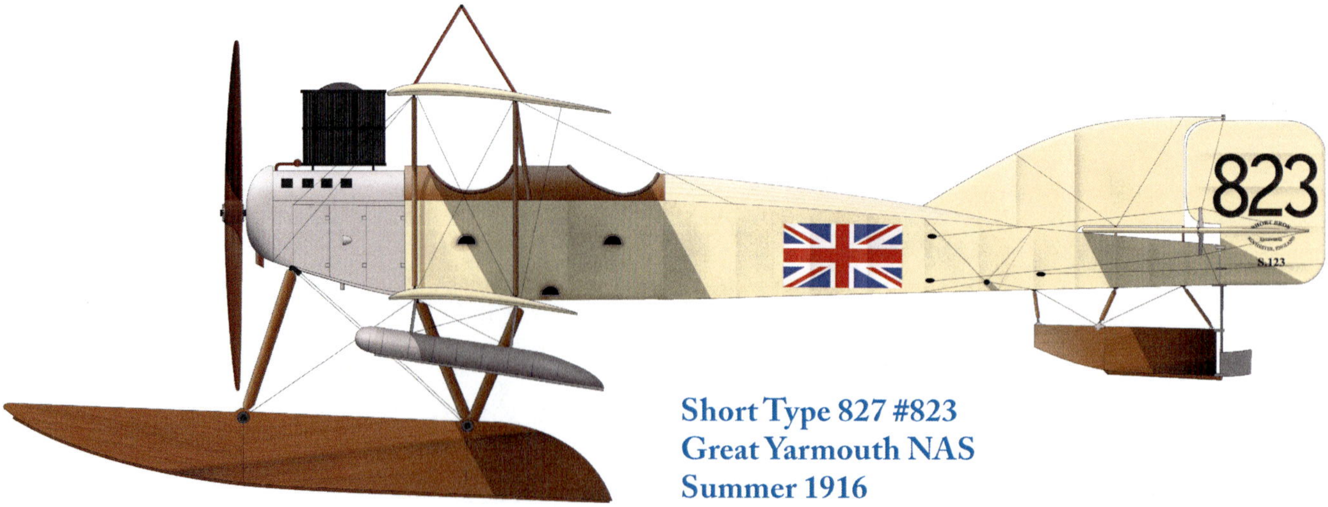

Short Type 827 #823
Great Yarmouth NAS
Summer 1916

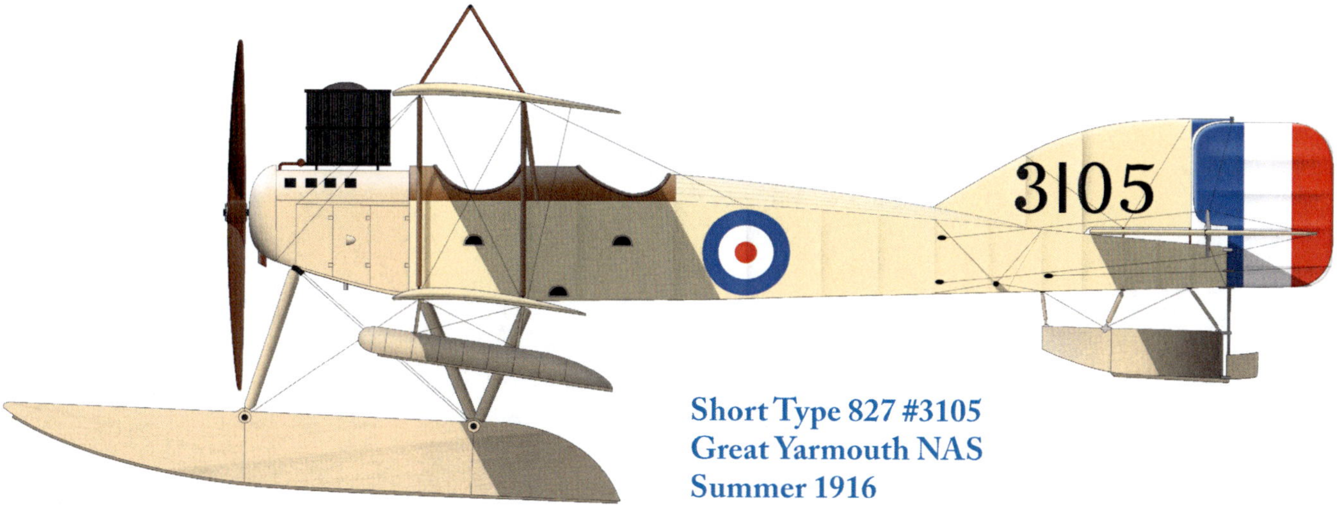

Short Type 827 #3105
Great Yarmouth NAS
Summer 1916

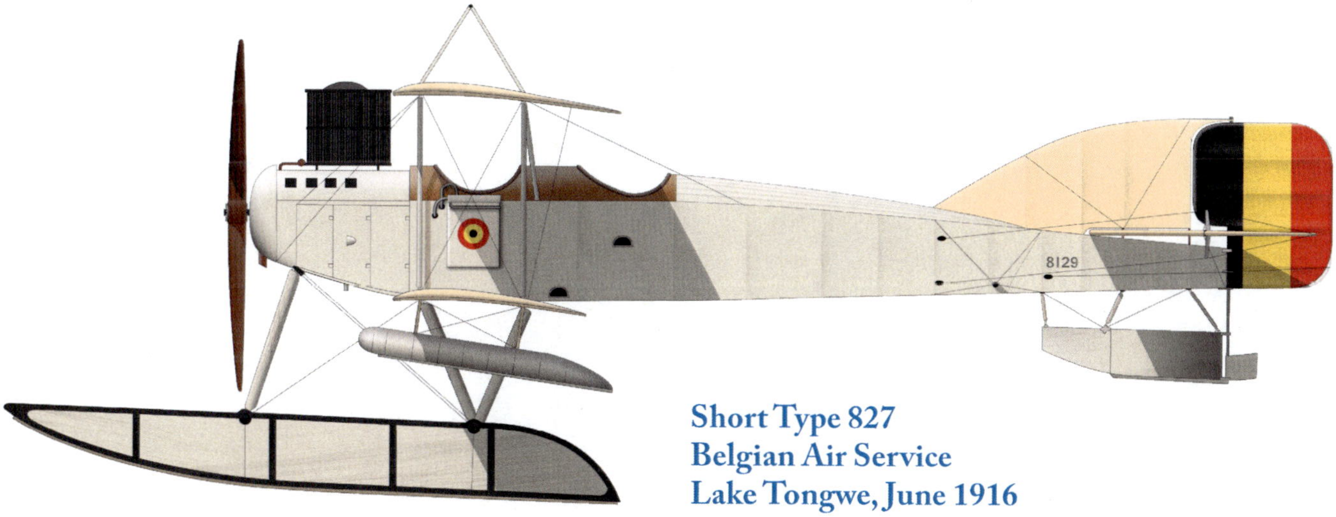

Short Type 827
Belgian Air Service
Lake Tongwe, June 1916

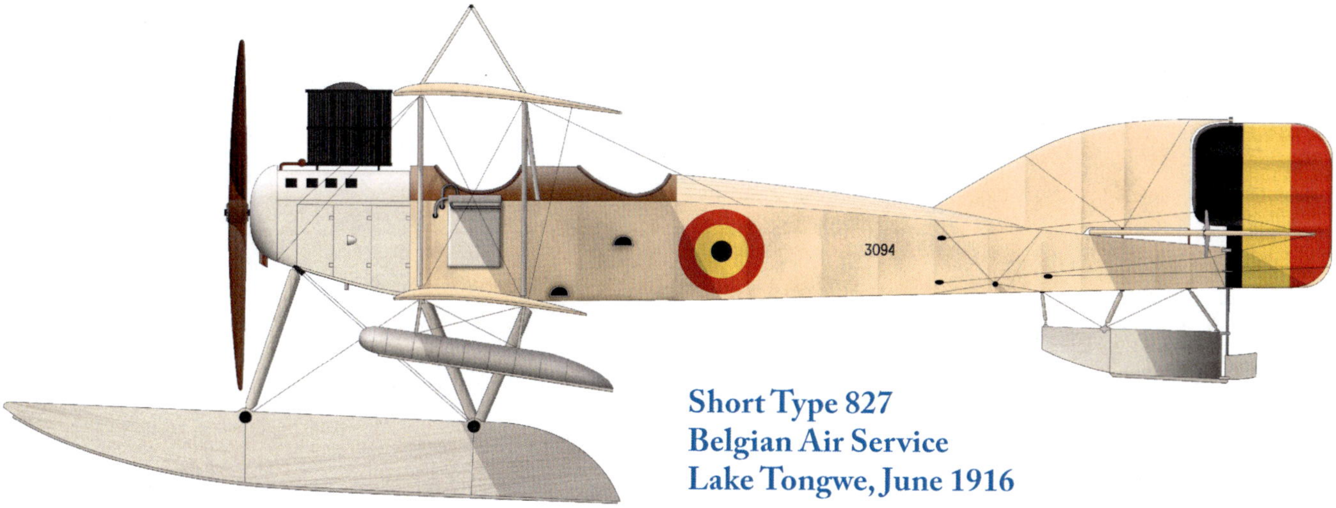

Short Type 827
Belgian Air Service
Lake Tongwe, June 1916

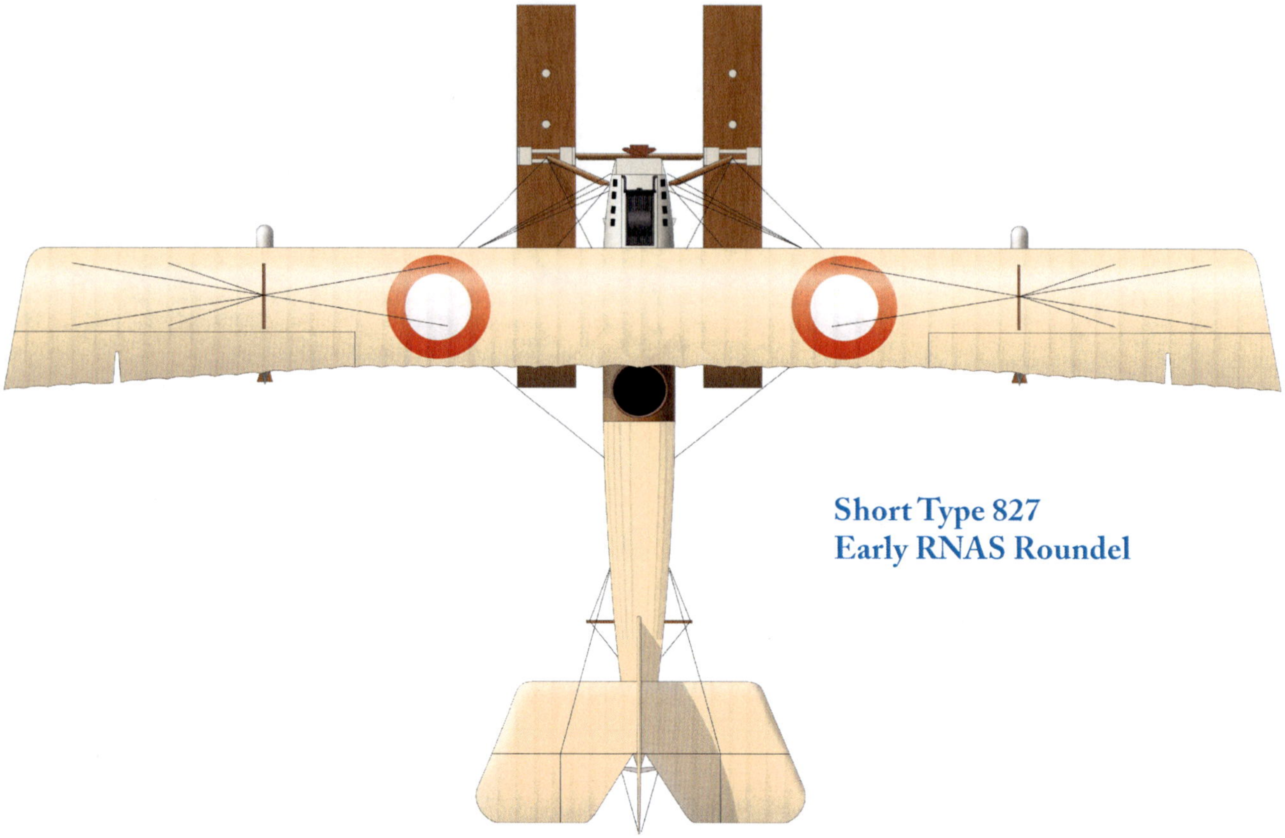

Short Type 827
Early RNAS Roundel

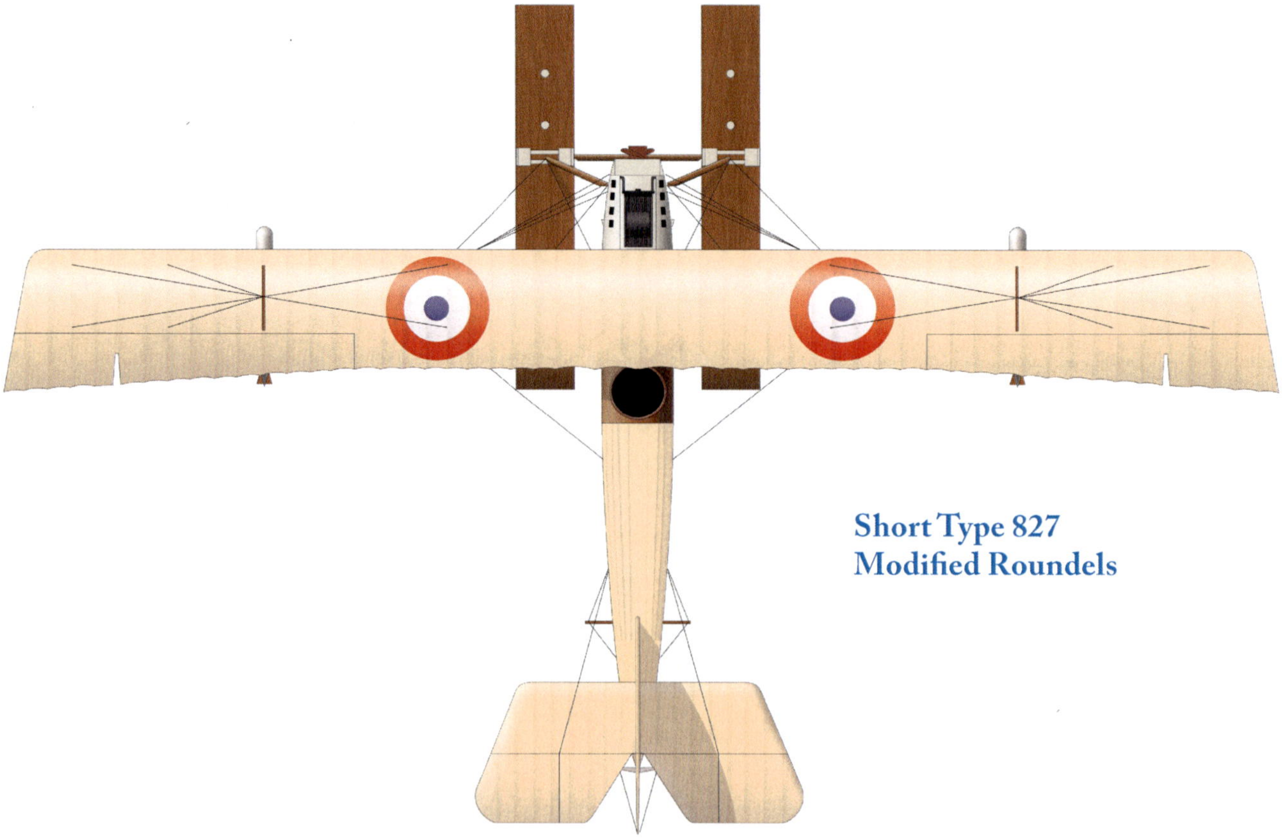

Short Type 827
Modified Roundels

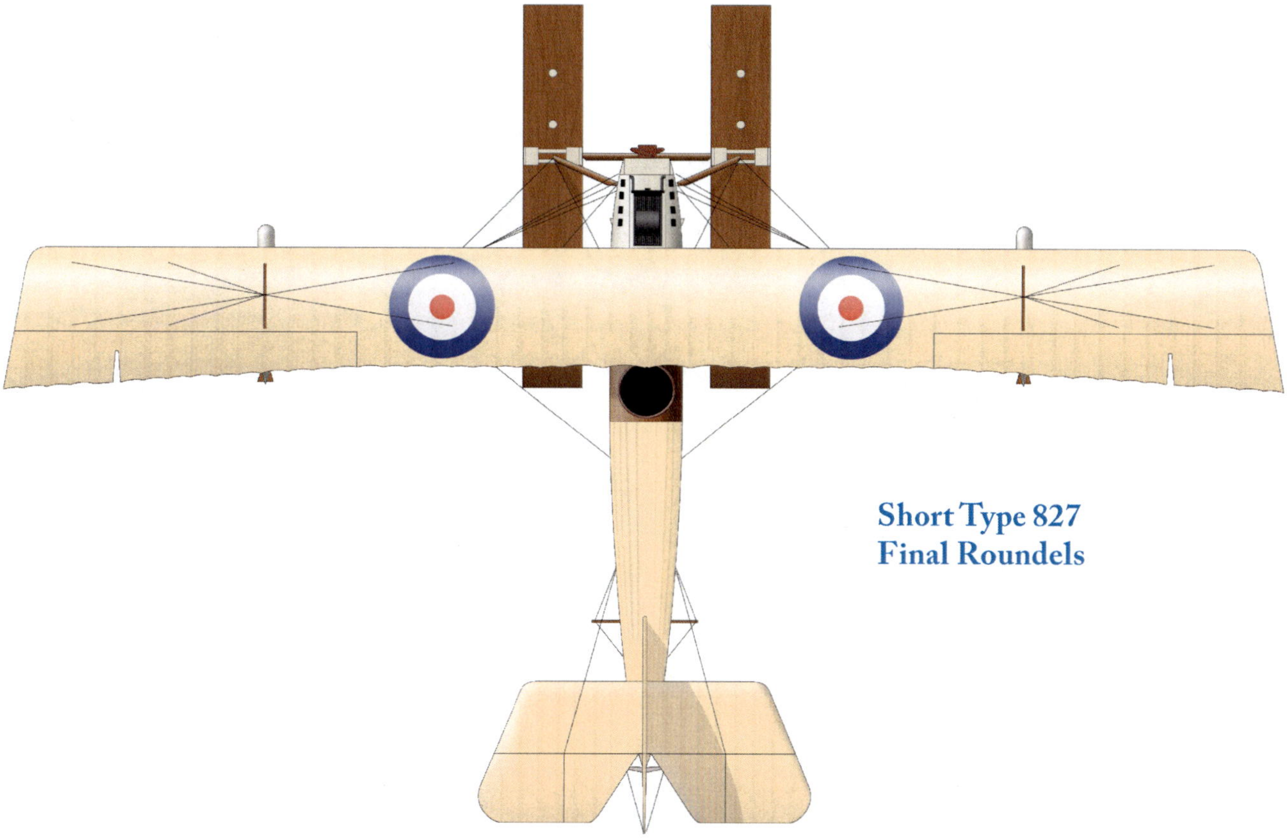

Short Type 827
Final Roundels

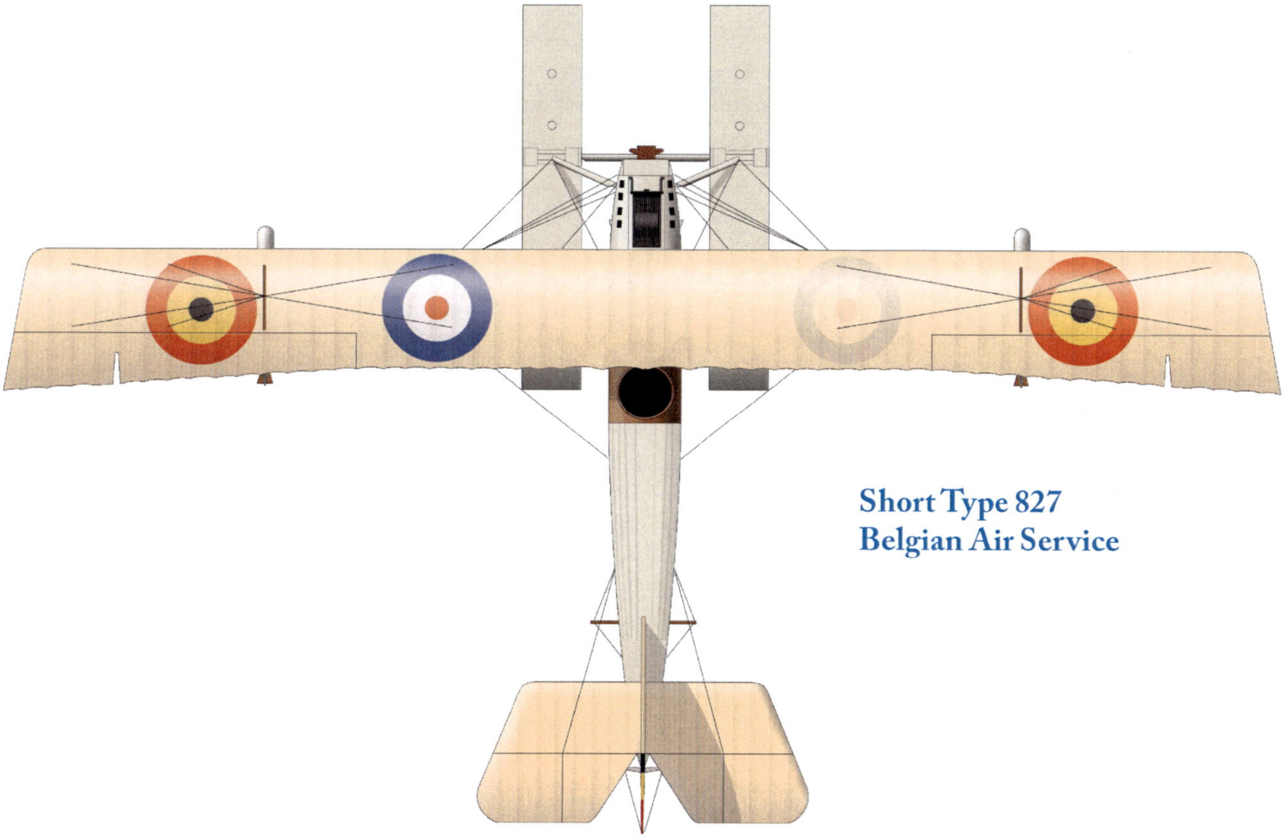

Short Type 827
Belgian Air Service

Chapter 9: The Short 830 Type

The origin of the Short 830 Type has been described with its twin, the 827 Type in Chapter 8. It was a two-bay biplane with the upper wing extensions braced by cables from kingposts above the outer set of interplane struts. Inversely tapered ailerons were fitted to the upper wing only and had straight trailing edges that extended inboard of the outer interplane strut, a distinguishing feature to identify the type from the Short 166. The interplane struts were of plain steel-tube. At least one 830 Type appeared with straight trailing edges to the mainplanes, unlike the usual Shorts' practice of a wire trailing edge that gave the types their typical scalloped appearance. This may have been associated with experiments on new aerofoil sections by the Admiralty Board of Invention and Research.[1]

The initial six were covered by two contacts (Nos 819 – 821 and Nos. 828 – 830). They were followed by contracts for twelve (Nos. 1335 – 1346) and ten (Nos. 9781 – 9790). Although only 28 Type 830 seaplanes were built, the type had an eventful service life. The first seaplane, No. 819, was described as having main spars of spruce, with ribs of three-ply with spruce capping. The undercarriage was of steel tube with spruce streamline covering. Floats were of spruce framing with three-ply covering. The fuselage was built of ash longerons with spruce cross beams and perpendiculars with wire cross bracing. Armament was described as *"Dover." Centre Section Circular Lewis Gun mounting*. Instruments fitted to No. 819 were an Inclinometer, Airspeed Indicator Patt No. 382, Aneroid Patt No. 172, Compass Patt 200, Cross Level No. 266, Revolution counter Elliott No. 114934, and Trip Clock.[2] Delivered to Grain on 14 December 1914, No. 819 was accepted on 12 January and delivered to HMS *Engadine* on the 15 January.

Short 830 No. 829 had taken off from Sheerness on the morning of 18 December 1914, for a flight along the coast. The crew was Sqn Ldr John William Sedden and leading mechanic R.L. Hartley. Due to problems, they had to alight at sea. The strong wind then blew the seaplane further out to sea.

The Short was found after about eight hours by the Norwegian ship, the SS *Orn*. The aircraft was picked up and brought back the Netherlands that same day. The crew was released and returned to Harwich on December 20.

The *Orn* moored in the Rijnhaven in Rotterdam. The aircraft was transported to the harbour park on a lighter and

Above & Facing Page: Batch No. 819 to 821 (S.119 to S.121) was the first batch of the 830 Type to be ordered. No. 819 appears to have an additional radiator under the nose. Shot down on 24 July 1916 by a German U-Boat North of Zeebrugge, it was later picked up by a German Torpedo Boat Destroyer. The crew of Flt Sub-Lt F.J. Bailey and Lt F.W. Madock, RNVR, were made PoWs. The seaplane is shown at Zeebrugge after its capture. Note how the ailerons hang down. The elevators were painted red/white/blue. (AHT AL0087-099 & 100)

placed under the management of the garrison commander with the inventory. The Short was then disassembled by the *Luchtvaartafdeling* (*LVA* – Aviation Department of the Netherland's Army) personnel and transported to the Navy Yard in Hellevoetsluis. The aircraft turned out to be too badly damaged to be repaired. The engine was taken over by the *LVA* for *f* 33000.[3]

Air Marshal Sir Ralph Sorley, KCB, OBE, DSC, DFC, has left a series of notes recalling his association with aircraft in his time serving with the RNAS. He learned to fly on a Short S.38 pusher and went on to fly a number of aeroplanes and flying boats during WWI.[4]

On joining HMS *Engadine* in early 1915 he flew a Short 830 Type Tractor floatplane with a 135-hp Salmson (Canton Unné) engine. *As usual, all one had was a few minutes flight as a passenger before being pushed off solo. … It was a really good aeroplane and handled easily and responsively for its size. The engine was on the heavy side for the power but was beautifully designed and made, and very reliable. In many hours of flying them I never had any engine trouble. They were difficult to 'unstick' with a full load in a calm but the floats stood up to a pretty pounding without breakage; if anything gave it was the chassis struts that bent. Once, with a heavy load, I careered a long way through choppy seas with water being thrown right over the engine when, in the end, the wooden propeller burst, which was not surprising.*

These aircraft were supposed to operate in open water from seaplane carrier ships, this *was highly optimistic as there was no weather reporting and conditions were either too rough or too calm for any certainty of take-off,* which he considered a highly inefficient way of conducting operations.

I once did a test, with a destroyer going ahead of me pumping oil on to troubled waters to see if by taking off along a path of oiled water we could find it noticeably calmer, but it did not seem to make much difference.

Flt Sub-Lt Sorley, RN, flew Short 830 seaplanes Nos. 819, 1336, 1340 and 1341. They were fitted with a Rouzet wireless set that *was mounted on top of the petrol tank and as it had a rotary spark gap revolving in the open and making a continuous arc, I always thought it more than dangerous.* These seaplanes carried no armament and were completely defenceless, *but they were capable of carrying a few bombs, weather permitting.*[5] The Log Book of No. 819 reveals that the aerial was enclosed in the wing. This was common on Short floatplanes.

No. 830, the machine that gave the Type is designation had a brief career with HMS Engadine from 14 January 1915, for on the 29th the *Machine completely wrecked except for main floats & engine. Mainmast of ship carried away and fell fore and aft along the fuselage.*[6]

Above: Nos. 820 and 821 with another seaplane in the background. The early national insignia and application of the serials to the rudder are noteworthy. No. 820 has a four-bladed airscrew while No. 821 has a two-bladed one. The ailerons do not hang down and appear to be interconnected. The pair had parallel careers, allocated to Grain, HMS *Empress*, and finally HMS *Ben-my-Chree* at the Dardanelles before the HMS *Ark Royal* and return to UK in April 1916 for deletion.

The Short 830 in Operation G

The object of Operation G on 4 July 1915, was to conduct a reconnaissance of Borkum and the Ems River, and also the possibility of engaging any Zeppelins encountered.7 The force required three or four light cruisers, 14 destroyers and two seaplane carriers, HMS Engadine and Riviera. If the weather and conditions were suitable on reaching Lat 53° 40' N, Long 5° 36' E, the floatplanes were to be hoisted out, the TBDs (Torpedo Boat Destroyers) would act as guard for the seaplane carriers while this was undertaken. The carriers would then proceed at speed to westward to the position where they would pick up the returning seaplanes. A division of TBDs would guard the carriers for this operation.

Three submarines were to be in the vicinity to pick up pilots from crippled seaplanes. Also. one submarine was to be positioned off Norderney Gat in case of an enemy ship coming out.[8]

Riviera and *Engadine* took part in Operation G. The idea was for the seaplanes to undertake a reconnaissance of Borkum but to drop two bombs in order to make the enemy think that the object of the attack was to bomb the facility.[9] *Reconnaissance is the principle idea of this operation, and a careful report of the enemy ships seen is to be made care being made to write down everything seen, with times and positions as accurately as possible.* The machines were, if possible, to avoid flying over Borkum due to its anti-aircraft defences and *not violate neutral territory if this can be helped*' Bombs were to be dropped on any enemy ships seen or other points of military importance.

If in danger of capture, every effort should be made to destroy the machine on the water. Revolvers and ammunition are to be thrown overboard before capture, also the bomb sights, charts, etc., and in fact as much done as possible to completely wreck the seaplane.

In the event of losing the Fleet, machines are to try and come down near Dutch territory or a neutral ship, but preferably Dutch territory, and disarm themselves. If possible, Pilots should fall into Dutch hands and claim the privilege of a shipwrecked mariner, i.e., they should not on any account admit being a prisoner of war liable to internment, but demand repatriation.[10]

Escorted by their accompanying surface craft of the Harwich Flotilla, the two aircraft carriers set off over the North Sea to a position about 30 miles off the Frisian Islands at dawn. The sea was calm enabling the seaplanes to take-off. *Rivera* hoisted out Short 830 floatplanes Nos. 1335, 1337, 1338, and 1339, all were on the water by 0240. Flt Lt Edmund Ivan Montfort Bird reported: -

On 4th July (1915) *I was hoisted out of H.M.S. Rivera in Seaplane 1339 and got off the water 5 minutes later and flew in an Easterly direction.*

On the way I saw 4 Zeppelins at short intervals of time going in the opposite directions.

I dropped two bombs to try and gain more height and then only managed to attain 4000 feet. After a half an hour from starting there were thick low lying clouds, well below my height, which prevented my seeing the land until I reached Juist which I could just recognize and here I turned and flew in a westerly direction, and I could just see a small portion of Borkum and also 6 destroyers quite close here.

Continuing to fly in a westerly direction where I expected to find H.M.S. Rivera but could see nothing of her.

At 5.45 A.M. I noticed that there was no more oil, and so

Above & Below: No. 1335 shows off the blue over-painting that some 830 Type seaplanes carried. Note the red ring early RNAs insignia on the wings, and the Shorts logo on the bottom of the rudder. Originally fitted with a 135-hp Canton-Unné radial, this was changed to the 140-hp model in early 1916.

Below: No. 1336 was another with the blue over-painting. It does not appear to have the red-ring markings on the upper wing.

Above: No. 1340 in the plain doped scheme. Note the unusual red ring markings on the upper wing. The pilot's handwheel can be seen in the original print. No. 1340 was with *Engadine* in July 1915; served at Dunkirk and Dover through late 1916, and was deleted on 9 April 1917.

I dropped the two remaining bombs to make certain of a safe landing, when I put in one gallon of oil out of a spare tin. After getting off the water again I (s)potted a Zeppelin quite close going in an easterly direction and about here I also potted (sic) *over a cruiser and some destroyers which I believed to be German.*

At 7.0 A.M. there was very little petrol showing in the gauge, and so sighting a Dutch trawler I landed near her, and they threw me a line which I made fast to the machine, whilst I attempted to destroy her, then I was taken on board the trawler and asked the captain to run the machine down, which he did sinking it almost at once.

The trawler took me to Muiden (Imuiden) *where I was taken to the Commanding officer to whom I gave my parole.At Muiden* (Imuiden) *I stayed with the British vice-counsul. On the 7th of July I heard that I was free and returned to England on the 8th.*[11]

The Dutch reported that at approximately 0900 hours a seaplane landed alongside the *YM175.* The pilot was allowed to board. He punched holes in the floats, drained the petrol tank and tried in vain to set fire to the plane with flares. At his request, the *YM175* overran the plane. After this, the *YM175* sailed to Ijmuiden, where skipper P 't Hart took the pilot to the police station. Flt Lt Bird was released as a shipwrecked mariner four days later.

What is surprising from the Dutch account is that Bird had an aerial encounter with *Oberletunant* Wolfgang von Gronau. The breakage of a fuel line caused Von Gronau to stop the fight. He then landed in the ground mist and managed to repair the line, after which he returned safely to his base at Borkum. Bird did not mention the fight with von Gronau in his report.[12]

Flt Lt H Stewart in No. 1337 had better luck. He saw three Zeppelins on his flight out and another that he thought was a Schute-Lanz *right ahead, climbing at a steep angle.* (This would have been the *SL 3.* The Schute-Lanz airship were built of wood unlike the Zeppelins with their metal frames.) Proceeding to Borkum, he noted four large destroyers and two small torpedo boats. Twelve torpedo boats were at anchor about three miles due south of Borkum in Randzel Gat. He flew over Knock Point dropping a bomb on the battery there. *Shortly afterwards a shell burst about 1000 feet above me, and this was the only fire encountered.* He then flew over the torpedo boats lying in Randzel Gat and dropped two bombs but was unable to ascertain what damage they had caused. He noted an enemy seaplane about 3,000 feet below him and the four Zeppelins that appeared to be *observing the movements of our own ships, came down to look for the ships under the clouds.* He sighted *Riveria* and dropping his last bomb; made a successful alighting at 0550 and was taken aboard.[13]

With regard to Flt Sub-Lt Gerald Edward Livock in

Above: A poor photograph of No. 1341 showing that it had the red ring RNAS markings with the Union flag on the fuselage. Flown by H.H. Gonyer at Dover on 19 May 1917, having been allocated there in April. Then to a training unit at Calshot in July 1917, and finally deleted the W/E 14 September 1917.

No. 1335, and Sqn Cdr F.E.T. Newlett in No. 1338, these machines returned to the ship after about half an hour due to their engines not running correctly or giving full power. This was the correct action under the circumstances.[14]

Livock wrote that he had been north on the *Engadine* for about a month when he was lent to *Rivera* for the operation and had to leave his own Short 830, No. 1336, behind on *Engadine.* He was given No. 1335 that was armed with four 16-lb bombs, a revolver and a Very pistol. *Considerations of weight did not allow an observer to be carried.* After managing to take-off successfully, he climbed and after reaching an altitude of 1,600 feet was over German territory.

Then the engine faltered. In a rare state of panic I turned round and steered back towards the ships with the engine pooping and banging, but luckily maintaining just sufficient power to keep me in the air.

I had no sooner settled on my course when, to my utter amazement, I found a huge zeppelin between me and the ships, about 500 feet higher than I was; then I saw another in the distance, and then another, until eventually there were four within a few miles of me. In my distraught imagination I could see myself surrounded by Zeppelins and being forced to fight to the end with a revolver in one hand and a signal pistol in the other, while my huge adversaries shot me to pieces with their machine guns. In spite of my panic, I did the correct thing. I more or less closed my eyes and flew straight under the nearest Zeppelin towards the ships, whose smoke was visible ahead. By the time he reached the ships, Livock was sweating with anxiety *and only thought of getting on to the water and back to the* Riviera *as quickly as possible. Sad to relate, I made a complete hash of the operation, for, after landing, I taxied to the ship's side much too fast, missed the crane hook and hit the ship's side with a splintering crash, which shattered my propeller. The handling party, however, heaved a lint to me, and I was hoisted on board.*[15]

Above: No. 1341 at a later stage with cockades to wings and fuselage and unusual tail markings. Note the tailplane and elevators are a dark colour on their upper surface. Curtiss H-12 Large America flying boats are in the background.

Stewart's performance greatly impressed his CO. In his report he drew attention to the fact that Stewart was a Direct Entry Officer without experience in navigation or very extensive experience in observation *such as a Naval Officer in his position would have.* Despite this Stewart performed an extended reconnaissance than ordered *and his report, without Naval training, appears to me to be very complete and concise.*[16]

At 0230 hours *Engadine* had launched Short No. 819 with Flt S/Lt Sorley as pilot, together with three Sopwith Schneider floatplanes as their part of Operation G. Engines were started up and warmed up while the aircraft floated on the water. No. 819 was having difficulty in starting her engine and CPO Mec Brooks was sent to assist. Having started the engine, he jumped overboard and was picked up in the dinghy. Sorley then began to taxi No. 819 but she was caught in the wash of a destroyer and split her propeller and she had to be hoisted in.[17] All the Sopwiths had trouble, their floats breaking up, one was salved and two were lost. All three pilots were rescued. Despite the failure, the raid did achieve its purpose in luring Zeppelins out. *Whilst waiting to pick up seaplanes "Rivera" and her escort were closely watched by two airships at first, and their number was subsequently increased to four, but the airships took care to keep well out of range of the Anti-Aircraft Guns.*[18]

A German submarine had reported Tyrwitt's force as they made their way across the North Sea, but the German Admiral, von Pohl, did not send a strong surface force, deciding to rely on aerial reconnaissance the following morning. Six airships were sent out with orders to bomb the seaplane carriers. *L 9* from Hage sighted the British force, but her captain, *Kapitänleitnant* Odo Loewe, could not attack as the airship would not go beyond 5,000 feet and there was no cloud cover. Loewe considered this altitude was within range of the British guns, and one of the light cruisers and four destroyers kept firing at him in an attempt to drive him off. At 0651 German time (GT), Lowe saw a seaplane land alongside one of the carriers and be hoisted aboard. This would have been Stewart or Livock.

L 6 found the enemy ships around 0400 GT This airship was forced up to 5,200 feet and moved to attack but the heavy anti-aircraft fire convinced her captain that this was too dangerous and he withdrew, following from a safe distance as the ships withdrew.

SL 3 and *L 11* also found the British force. With four

Above: This 830 Type has the same type of marking around the rear fuselage. There are cockades on the wings but none on the fuselage. Note the Lewis gun armament through the hole in the upper centre section. The instrument panel is external of the cockpit in front of the pilot.

airships in sight, three brand-new Sopwith Schneider floatplanes were hoisted out by *Engadine.* As noted above, the sea, although only moderate, was too rough for the seaplanes and their floats broke up. Two sank but the pilots were rescued.[19] The operation was successful in that it lured the Zeppelins out, but the Sopwith Schneiders failed in mild North Sea condition. Even if they had been able to get off the water, their performance would have made chances of an interception unlikely.

The British official history, *The War in the Air,* states that the repeated failures of these expeditions caused many naval commanders to lose faith in the RNAS and the use of aircraft carriers. The experiences of the Commander in Chief with the *Campania* had done nothing to allay these fears. The seaplanes had great trouble in hoisting out and broke their floats when set upon the water. It was appreciated that a deck that allowed the seaplanes to take-off from the ship was required.[20]

HMS *Empress*

When HMS *Empress* was being readied for sea at Liverpool, from 9 May to 15 July 1915, by the addition of a steel aft hangar to fit her for service as a Floating Seaplane Base, the Vice-Admiral Commanding the Coast of Ireland was informed that when ready for sea, the *Empress* was to sail to Queenstown[21] *to act as a floating seaplane base, with headquarters in that port, her seaplanes being used in conjunction with the Patrol Vessels under your orders for the detection of enemy submarines. ... The EMPRESS will carry 4 "Short" Seaplanes, with 135 h.p. engines. These machines are well suited for patrol work and are the best available for the purpose required. Their speed and climb are moderate but they carry a passenger and four hours fuel.*[22] The seaplanes referred to would have been the Short 830 Type. It was proposed that the seaplanes operate from temporary stations along the coast, and *Empress* was to patrol at sea and launch her seaplanes as required. The ship was usually unescorted on these occasions.

A different type of event involved the aircrew attached to HMS *Empress* in September 1915. On the 4th, No. 1344 with Flt S/Lt Dunn as pilot, with Leading Mech Stevens patrolled from Galley Head to 15 miles south of Fastnet from 1025 to 1210 and reported 'Patrol Clear.'

No. 1342 with Flt Lt Field and observer, Flt S/Lt Lott left at 1605 and returned at 1745, having patrolled 15 miles on either side of the course of the ship, and reported Area Clear. Then at 2015 the ship received an SOS from the SS *Hesperian* that had been torpedoed. Rushing to the scene, *Empress* was the first ship to arrive and found about 12 boats full up with passengers and crew. These were picked up with some difficulty. The ship was ordered to return to Queenstown with the 430 survivors that they had picked

Above: No. 1346 showing the early colour scheme. Union flag on fuselage, the rudder contains the serial number on white field above the Short logo and the Short c/n on the bottom of the rudder. All fabric covered surfaces are clear doped with the ply and metal panels presumable grey. Floats appear to be the same colours the fabric and may have had fabric over their plywood. Note the number of crew required to launch a seaplane.

Above: No. 1346 at a later date with serial now on the fin and rudder stripes and cockade national markings. Photographed at Great Yarmouth, probably during the winter of 1915–1916, note the snow on the ground. It appears to have spent most of its life between Dunkirk and Dover.

Above: This Short 830 appear to have had some trouble. The floats are loaded on a truck, completely separate from the fuselage. This machine has the red ring wing markings with a white outline, and the blue over-painting. Serial number unfortunately unknown.

up. They arrived at the Deep Water Quay at 0342 on the 5th. The ship left again for the wreck with volunteers and crew at 0925.[23] It is sometimes not remembered that the RNAS officers were engaged in other naval activities as well as well as those involved with flying.

In December 1915, it was decided to change *Empress'* seaplane complement *to two 225 HP Sunbeam Short and four Converted Schneider Seaplanes.* As it was not possible to erect all the machines on board, two of the small seaplanes would have to remain in their cases. Short 184 Nos. 8048 and 8381, and Sopwith Modified Schneider seaplanes Nos. 3772 to 3775 were allocated to the ship. No. 3772 was at Calshot and all the other machines were to be assembled at Calshot. The Schneiders were to be sent direct from Sopwiths, except No. 3772.[24]

Empress left Plymouth for Gibraltar on 9 January 1916. In February she was reported to be in Genoa for repairs that would take to about 10 March to finalise.[25]

Dunkirk Anti-Submarine Patrols

No. 819 conducted *Expedition along coast from Dunkirk* flights from 11 to 16 February 1915. What these flights were to achieve has not been ascertained to date.[26]

30 April 1916. No. 1338, with crew of Flt Sub-Lt Cowley and AM Fryer, while on anti-sub patrol about seven miles out to sea from Zeebrugge, sighted two German seaplanes and engaged them. The HA replied, but Fryer's the gun jammed three times spoiling shooting. The HA retreated to Zeebrugge and Cowley continued his patrol. At 1120 he sighted a submarine by buoy No. 9 and came down to 700 feet to investigate, deciding the submarine was friendly, but when alongside, was fired at by a machine gun. He promptly sheared off and by firing a series of Very's lights called the attention of our destroyers that came up, but the submarine disappeared. They searched for the submarine without success.[27]

No. 1338 had more close encounters with the enemy. Leaving Dunkirk on 1 June 1916, at 1025 Flt Sub-Lt Tooke (pilot) was to look for missing Short 184 No. 8346. Nothing was seen of this machine but a hostile submarine was sighted 5 miles magnetic north of Zeebrugge steering south at 1102. At 1104 the submarine was again seen. It turned slowly and headed north. The submarine was proceeding slowly and half blown.[28]

Three days later, Flt Sub-Lt Bailey with Sub-Lt Slatter as observer, left Dunkirk at 0739 for a patrol. They sighted three German TBDs eight miles off Zeebrugge that opened fire on the seaplane without effect. The position enemy was reported to nearest British destroyers who fired on the enemy, one shell being seen to burst above centre of the hostile TBDs.[29]

Flt Sub-Lt Fisher with Observer Sub-Lt Swanston, RNVR, carried out a climbing trial on No. 1346 on 3 July, 1916. This 140-hp Canton-Unné powered Short reached 7,500 feet in one hour and *would have climbed to 9,000 ft. had not one cylinder misfired.*[30]

On 24 July, No. 819 from Dunkirk attacked a surfaced U-boat some 12 miles off Zeebrugge. The U-boat did not try to dive but fired at the seaplane hitting the engine and forcing it to alight. The crew of Flt Lt J.F. Bailey and Flt Lt F.W. Mardock, still had sufficient engine power to taxi and headed towards the Netherlands but were overtaken by a torpedo boat and taken prisoner.[31]

Seaplane No. 1341, with pilot Flt Sub-Lt Durston and observer Sub-Lt Tidey, left Dunkirk at 1445 on a search patrol for missing Sopwith Baby No. 8138 on 16 November 1916. The Baby had come down with engine trouble. The

Above: On many photographs of Short floatplanes being hoisted with the wings folded, there are men on the floats. This is thought to have been necessary to balance the machine on the hoist.

Facing Page: *135 Salmson* was the caption to this photograph of No. 1341 (so marked on the nose of the float). This clear doped machine has the red ring marking with the white outline to the upper wings.

Short 830 was one of seven search patrols sent out to find the missing seaplane.

At 1420 they sighted two hostile seaplanes between Nieuport and Ostende, and *engaged in a running fight for 15 minutes, when hostile machines turned back in direction of Ostende. When distant a quarter mile the leader of the hostile machines fired white stars, and immediately after attempted to cut our machine off and drive her to Eastward. The other hostile machine meanwhile manoeuvred into position above and behind our machine and opened fire. Our seaplane then headed directly for the leading hostile machine and when distant 200 yards fired a few short bursts at him. He immediately turned back and took up position with the other machine, above and behind ours, and for 15 minutes a running fight was carried on at a range of 100-150 yards. Tracers were observed entering the fuselages of both enemy machines, and the leader appeared to have been hit badly towards the end of the fight, as while in excellent position for firing on our machine he suddenly went into a very steep left-hand bank, dived several hundred feet, flattened out and headed for Ostende. The remaining machine also turned back about two minutes after. These machines have a speed of about 80-85 knots, but the pilots did not seem to care to execute many manoeuvres, and certainly did not make the most of their advantages in speed over the Short Seaplane, which at the most was doing 63 knots.*[32]

P.S. Fisher from Dunkirk flew No. 1341 on 27 February 1917, and recorded – *Joy Ride. Control from front seat.*[33] No. 1341 was at Dunkirk until April when it was at Dover, then at Calshot as a trainer on 5 July 1917. Documentation has not been found to determine whether this was a one-off modification at Dunkirk or if all the obsolescent 830 Types were so modified.

The venerable first 830 Type, No. 819 was operating from Dunkirk from 1 July 1916. The *Machine nose heavy. Anhedral removed from tail plane by lowering front chocks ¼". Fuselage opened up & examined. Wires cleaned of paint. Lead of aileron control wires altered necessitating new shells being made. Loose wires renewed.*[34] After this overhaul the seaplane was ready for testing on the 11th and commenced patrolling on the 17th. Patrols were recorded in the Log Book as *Patrol* and *Fighting Patrol.* The last entry in the Log Book was for 24 July 1916: - *Left on patrol. Did not return. From signal by pigeon carrier, machine shot down by Hostile Submarine and afterwards picked up by hostile TBD. Pilot and passenger taken prisoners.* Flt Sub-Lt Bailey and SL Mardock, RNVR, were both made PoW.

Above & Below: Landing No. 820 at Imbros from HMS *Ben-my-Chree*. The locals have all pitched in to help. Note what appears to be some tools on the fuselage side under the pilot's cockpit.

Home Defence

In 1915 German seaplanes made attacks on shipping in the Channel, Thames Estuary and close to British ports. The SS Blonde reported that on 15 March, *we observed a water plane approaching us from the North-ward. We at first thought it was English. Soon found it was a hostile craft. She circled round us & first dropped two bombs which fell within 20 ft. of the stern. Then sweeping around again, she dropped two more bombs which fell about the same distance ahead of us.*

Then again sweeping round came up astern and dropped one close on our starboard side. Five in all.[35]

Actions like this led to the response by the Naval authorities of sending aircraft and seaplanes out on receiving a message that enemy aircraft had been seen. A report in April 1915, noted that Short No. 820 left Grain at 1340 in response to a report of enemy machines had been seen, and went out a considerable distance beyond Galloper, but the enemy had retired. Returned at 1725 having been in the air for 3:05 hours.[36]

Although they did not have the performance to be

Above & Below: This photograph of No. 9782, from the last batch of Short 830 seaplanes, shows the cut-out in the upper centre section, compass mounted in front of pilot's cockpit, and the Shorts logo on the rear fuselage. The colours are unknown but the fabric does not appear to be clear doped. Most of its service was at Grain with the Design and Experimental Flights. In June 1917 it was at Calshot, then Lee on-Solent, presumably on training duties. Deleted 8 November 1917, due to fair wear and tear. No. 9782 at Lee-on-Solent is definitely in a late P.C. dope scheme. Note the camouflage on the canvas hangars.

Short 830 Serial Allocation			
Serial	**Contract**	**c/n**	**Notes**
819 – 821	CP57105/14/X.18826	S.119 – S.121	135-hp Canton-Unné.
828 – 830	CP57105/14	S.116 – S.118	135-hp Canton-Unné.
1335 – 1346	CP37639/15	S.151 – S.162	135-hp Canton-Unné, most refit-ted with 140-hp later.
9781 – 9790	CP118670/16	S.301 – S.310	Hybrid school machine.

effective, seaplanes were sent out whenever a bombing raid was taking place on England. On the afternoon of 25 May 1917, 23 Gotha G.IV bombers of *Kagohl* 3 left their base to attack the UK. 21 succeeded in reaching their objective. On receiving a *Warning*, Dover sent Short 830 No. 1337 out on patrol at 0525 crewed by Flt Sub-Lt Leslie C Pincott with CPO Hanna in the rear. The machine passed over Folkstone a few minutes before the hostile aircraft attacked the town. No hostile aircraft or shipping was sighted by the pair.

At 1815 No. 1337 was sent out with Pincott again as pilot, and with CPO Hanna as observer.[37] While flying out of the harbor the enemy machines appeared overhead and when some three miles from Dover, they sighted some ten large, hostile machines overhead. Six bombs were dropped into the sea within 500 yards of their machine on the port side. Pincott was unable to attain the necessary height to attack the enemy machines, and by the time he had reached Calais the enemy machines had disappeared. He returned to Dover at 1850 in the evening.

Dover complained that *The only "fighter" seaplane in commission on this station was out of action through engine trouble and the Engineers were working on this machine at the time of the hostile attack.*

Seaplane No.1337 was the best machine available (No.8102 being on duty at Dunkerque) but this machine, being of an old design, could not attain the altitude necessary to attack the hostile craft which were flying at about 14,000 feet.[38]

The landplanes from Dover did little better than the solitary seaplane, having limited success. Sopwith Pup No. 3691 engaged a large enemy two-seater mid-Channel, and reported that black smoke was seen coming from the enemy machine when it *was attacked from above and behind by two more Hostile Aircraft and after breaking away from these was forced by engine trouble to return.*[39]

The raid killed 95 persons, injured 195 and caused £19,405 damage. 77 sorties were flown by the RNAS and RFC. One Gotha was shot down and another crashed near Bruges.[40]

No. 1337 was indeed an *old design*, being delivered to Grain on 19 May 1915, and served on HMS *Riviera* to June, then at Dunkerque and Dover. It was rebuilt after a chequered career in which it *collided with breakwater while being towed in causing damage to elevators, tail plane & pontoons, main plane* on 10 January 1916[41]; having suffered a forced alighting in fog, collided with a breakwater, in addition to the usual accidents that happened to operational seaplanes. Back in service by May 1917, a forced alighting due to engine failure on 18 August 1917, led to the machine capsizing. It was towed to Ramsgate and deleted on the 22nd.

S.302 to S.310 – The Hybrid Admiralty 830 Type

Shorts received an order in July 1916 for ten Short Admiralty 830 Type seaplanes. These school machines were listed under that designation but were a hybrid design with the wings and fuselage of the 166 Type, but with the observer in the front cockpit and straight-edged ailerons of the 830 Type. Serials were 9781 – 9790 (S.301 – S.310). These school machines had a Salmson Canton-Unné radial engine of 140-hp. The first, No. 9781, was delivered to Grain on 4 November 1916.

In June 1917, Flt Cmdr Sitwell, of South Shields, reported that he had only three machines for patrol work and only one was of any use. This was 240 Sunbeam Short 184 No. 9066 *which gave every satisfaction and was flying very well. It was used continuously for patrols.* No. 8007, a very old 225 Sunbeam Short 184 *was not now good enough for sending out on distant patrols.* It was used mainly as a training machine for practice flights. His third machine was No. 9784 a 140 Canton Unné Short Type 830 which *was of no use whatever as it would not get off the water with pilot only, let alone with passenger and bombs, etc.*[42] Since the type was meant to be a school machine it is interesting that it was tried to be used for patrol work. J.M. Bruce commented that *the fitting of bomb racks indicated an excess of optimism.*[43]

In July that year, the CO of South Shields said that the *two 150 Canton Unne Shorts which were of no use to him* could be sent to Calshot for school work.[44]

By the time the Short 830 seaplanes were entering service in November 1916, the 184 Type was in service and dominated the seaplane position in all roles, with the exception of that of torpedo launching, until the end of the war.

Referring to Great Yarmouth station in late 1915, an unnamed officer wrote that the complement of aircraft was about 15 of various types – seaplanes and aeroplanes. *For the most part they were good of their type and considering all things at that time they were reliable. Of the seaplanes, the Canton Unné Short was a wonderful machine. It always started when required, and always got off the water quicker than any other type of Short, and that usually with a greater load than it was ever intended to carry. So far as I can remember, it never let any one*

Specifications Short 830					
Source	**1**	**2**	**3**	**4**	**5**
Span upper		53 ft 11 in	53 ft 11 in	52 ft	
Span lower			40 ft	39 ft	
Length		35 ft 3 in	35 ft 3 in	36 ft	
Height			13 ft 6 in	12 ft 6 in	
Chord			5 ft		
Gap			5 ft 6 in		
Stagger			0°		
Dihedral			0°		
Wing Area, ft^2	506	506			
Empty Weight		2,624 lb.	2,622 lb.		
Military load			0		
Crew Weight			360 lb.		
Fuel & oil			342 lb.		
Loaded Wt, lb.		3,324 lb.	3,324 lb.		
Fuel (Main), gal.				38	30
Fuel (Gravity) gal.				15	15
Oil, gal.				6½	6
Max speed	61 kts/2000 ft	70 mph	70 mph/2000 ft	50 kts	50 kts
Alighting					5 kts
Climb to 1,000 ft.					6½ min
2,000 ft	10.4 min		10:25 min		10 min
3,000 ft				23 min	27.3 min
Endurance	3½ hrs	3.5 hrs	3½ hrs		1 hr 34 min
Engine	140 Can:Unne		135-hp Salmson		135H.P. Canton Unne No. 729

Notes: 1. *Including King Post on top plane.*
Source: 1. Chart "Performance of British Seaplanes 1917 – Dec 31, 1918." Date Test Report.17.04.1917. TNA AIR1/708/27/11/02.
2. Barnes, C.H. *Shorts Aircraft since 1900*, Putnam, UK, 1967.
3. J.M. Bruce Data.
4. Log Book for No. 819. TNA AIR1/2064/204/412/819.
5. Log Book for No. 1337. TNA AIR1/2065/204/412/1135
The 4 bladed prop AD No.187, 2590 pitch/1680 dia LH for the B.T. Short 830 was cancelled by Technical Memorandum No.81 of 19 September 1916. (Source: *Technical Memoranda*, 1916. Air Department No.19, 01.01.1917. AIR1/2097/207/15/3.)

down, which is a rather marvellous record.[45] The identity of this Canton Unnê Short is unknown but it shows how well the type was regarded.

End Notes

1. Barnes, C.H. *Shorts Aircraft since 1900*, Putnam, UK, 1967. P.105.
2. Log Book of No. 189. TNA AIR1/204/412/819.
3. Gerdessen, F & Geldhof, N. *De Internneringen van vliegtuigen tijdens De Groote Oorlog*, Uitgeveri Geromy bv, Netherlands, 2018. P.234.
4. Sorley, R. 'Sorties with Sorley.' *Cross & Cockade International*, Vol. 37 No. 2 & 3 UK, 2006. P.120 & 186. Sorley is best remembered for being largely responsible for the adoption of the Hurricane and Spitfire fighters by the RAF in time for WWII.
5. Sorley, R. "Sorties with Sorley", *Op Cit.*, P.126.

6. Log Book of No. 830 with Log Book of No. 189. TNA AIR1/204/412/849.
7. Robinson, D. states that the primary object of the attack was to lure the Zeppelins out where they would be attacked by the Sopwith seaplanes. *The Zeppelin in Combat*, Schiffer Publication, USA, 1994. P.105–110.
8. *Orders for Operation "G"*, from R.Y. Tyrwhitt, Commodore (T), HMS *Arethusa*, to CO of HM Ships concerned, 01.07.1915. UKNA AIR1/148/15/88.
9. Livock writes that the object was to bomb the wireless stations at Emden and Nordeich. Livock, G.E. *Op Cit*. P.26.
10. *Orders for Seaplanes*, issued by Lt Cdr E.D.M. Robertson (in command), HMS Riviera, 03.07.1915. UKNA AIR1/148/15/88.
11. Flt Lt E.I.M. Bird to Director of Air Dept, Admiralty, 08.07.1915. TNA AIR1/148/15/88.
12. Gerdessen, F & Geldhof, N. Op Cit. P.238.
13. Flt Lt H Stewart's Report to CO, *Riveria*, 04.07.1915. AIR1/148/15/88. The Log Book of No. 1337 has the one word – *Emden* – in the remarks column. The time in the air was recorded as 3 hrs 5. TNA AIR1/2065/204/412/1337.
14. Livock recorded that he was given Short Type 830 No. 1335, a pistol, a Verys pistol and carried four 16-lb bombs. Livock served from 1914 to 1945, retiring as Group Captain, G.E. Livock, DFC, AFC. He was second in command of the Far East Flight that flew to Supermarine Southamptons to Australia in 1927. His autobiography, *To the Ends of the Air*, was published by the IWM in 1973, and gives an overview of a remarkable flying career.
15. Livock, G.E. *To the ends of the Air*. IWM/HMSO, UK, 1973. P.27.
16. Lt Cdr XXX, *Riveria*, to Commodore (T), Arethusa, 05.07.1915. TNA AIR1/148/15/88.
17. In No. 819's Log Book Sorley only recorded -.20 hours duration and left the Remarks column empty.
18. *Ibid.*
19. The German airship side of this encounter is from Robinson, D. *Op Cit*. P.87–89.
20. Jones, H.A. *The War in the Air,* Vol.2, P.361.
21. Now Cobh in Ireland.
22. To Vice-Admr Commanding Coast of Ireland. Secret & Immediate, 23.07.1915. UKNA AIR1/436/15/284/1.
23. Lt F.W. Bowhill, in Command *Empress* to Vice Admiral Commanding HM Ships & Vessels, Queenstown. 06.09.1915. TNA AIR1/436/15/284/1.
24. "Seaplane outfit of "*EMPRESS*". Admiralty Air Dept. 13.12.1915. Short Nos. 8344 was substituted instead of No. 8048. TNA AIR1/436/15/284/1.
25. *Ibid.*
26. Log Book of No. 189. TNA AIR1/204/412/819.
27. RNAS Daily Communique Vol. 1. Apr to Dec 1916. TNA AIR1/39/15/9/4.
28. TNA AIR1/47/15/9/27. No. 8346 was lost on 1 June 1916,
29. RNAS Daily Communique Vol. 1. Apr to Dec 1916. TNA AIR1/39/15/9/4.
30. 'Dunkirk Seaplanes. Operations carried out between 1 July – 15 July 1916.' TNA AIR1/39/15/9/4.
31. Hobbs, D. *The Royal Navy's Air Service in the Great War*, Seaforth, UK, 2017. P. 196, and Henshaw, T. *The Sky Their Battlefield II*, Febuti Books, UK, 2014. P.46. Schmeelke, M. *Zeebrugge Naval Air Station Flanders I 1914–1918*, Aeronaut Books, USA. 2018. P.101 for the German version of events.
32. 'Report of Operations Dunkerque Seaplane Station. 6.00 p.m. 15 November 1916 to 6.00 p.m. 16 November 1916.' TNA AIR1/47/15/9/27.
33. Pilot's Flying Log Book P.S. Fisher. FAA Museum 2013/148/0001.
34. Log Book of No. 189. TNA AIR1/204/412/819.
35. A/S "BLONDE" at Sea. 15.13.1915. TNA AIR1/149/15/102.
36. 'Copy of Report to COS as to action taken against German aircraft.' 12.04.1915. TNA AIR1/149/15/102.
37. There is only one entry in the Log Book of No.1337 for 25.05.1917, for 40 minutes. TNA AIR1/2065/204/412/1337.
38. Sqn Cmdr V Gaskell Blackburn, CO Dover, to Wing Cmdr Lambe, HQ, RNAS, Dunkerque, 31 May 1917. TNA AIR1/637/17/122/152.
39. Co, RN Aeroplane Station, Dover to Senior Officer, Dover. 3 June 1917. TNA AIR1/637/17/122/152.
40. Cole, C & Cheesman, E.F. *The Air Defence of Britain 1914–1918*, Putnam, UK. 1984. P.232.
41. Log Book of No. 1337. AIR1/2065/204/412/1337.
42. Minutes of the 4th RN AD Progress Committee Meeting of 5 June 1917. TNA AIR1/270/15/226/113.
43. Bruce, J.M. "The Short Seaplanes" Part 3, *Flight*, 28 December 1956. P.1002.
44. Minutes of the 8th RN AD Progress Committee Meeting of 3 July 1917. TNA AIR1/270/15/226/113.
45. Snowden Gamble, C.F. *The Story of a North Sea Air Station*. Oxford-London University, UK. 1928. P.139.

Appendix 1: Colours and Markings of RNAS Seaplanes

Above: *Seaplanes at Kephalo.* Short 166 Types in the Aegean, painted on site by Norman Wilkinson, CBE, and published in *The Dardanelles 1915.* (Longmans & Co, UK, 1915). The blue-grey overpainting of the Short floatplanes as noted in the text is shown here. The few paintings that were made on site during WWI provide a good indication as to the actual colours carried by the aircraft during the conflict.

Colours and markings of British aircraft during WWI is a subject fraught with controversy. Even the colours of Protective Coverings P.C.10 and P.C.12 are still debated. The colours of RNAS aircraft are evenly more poorly covered than that of the RFC. The following notes are what can be learned from contemporary documents relating to RNAS doping practices, although the author acknowledges that there is still much to be uncovered.[1]

As early as August 1914, the planes of Sopwith No. 880 had been *doped with a blotchy blue colour as this was found to be a very inconspicuous colour for a seaplane flying close to the surface, a rather greyer colouring may prove more generally useful.*[2] A later report noted that No. 880's *blue colouring was too dark and that the bluish grey of* (Borel floatplane) *No. 84 is a better general colour.*[3] The bluish grey applied to early Short seaplanes would have been derived from these tests.

Short 830 No. 819 was delivered with the unbleached Irish linen fabric surfaces being given *three coats of Cellon and one Varnish (Elastic).*[4] Wight 840 Type No. 832's Log Book only contains the information that Emaillite dope was used.[5] Fabric surfaces were doped and varnished as a matter of course.

Temporary Memorandum No. 48 from 1915 recorded that certain firms had been found employing unsuitable dope on their aeroplane and seaplane's fabric. All firms were ordered to employ one of the three following dopes for their fabric in future:

(1). Cellon dope.
(2). Emaillite dope.
(3). Bi-nitro Cellulose Dope (as used by Messers Short Brothers).

The following general rules in connection with fabric have been laid down for the guidance of firms: -

(1). The fabric must not be put on the wings too taught, otherwise it is found that on doping the fabric it may shrink to such an extent to deform the wing.

(2). All joints in the fabric are to be of the double balloon seam type.

(3). The parts of the wing woodwork which come in contact with the fabric are not to be varnished, but are to be left bare, so as to enable the fabric to stick on to the woodwork when doping.

The dope is to be put on in the following manner in the case of Cellon and Emaillite: -

First coat: 2 parts dope, 1 part acetone.

Second coat: 4 parts dope, 1 part acetone, then 4 coats, or if necessary 6, of plain dope.

After the dope has thoroughly dries a doped varnish should be employed to finally make the wing waterproof, or else the following solution should be used: -

Take one part of best boat varnish and mix with two parts of petrol. This is spread over the wing with a brush, but brushing must be continued till the varnish becomes tacky. Each contractor ahs been ordered to send out a series of machines treated in this manner, and Commanding Officers are to report immediately should they find the doping on any of these unsatisfactory.

The attention of all Commanding Officers is drawn to the above, and when forwarding reports concerning machines they are to give the following details:

Type of the machine.

Constructor.

- *No. of machine.*
- *Date received at station.*
- *Number of hours flight in air.*[6]

In July 1916 the Air Department produced Technical Memorandum No. 59 that set out the doping practices for RNAS aircraft. Only non-poisonous dopes were to be used. For machines of 1st Class importance, they were:

Emallite No.11a and No.11c.

Cellon No. N.P.2.

Raftite.

The poisonous dopes containing tetrachlorethene were not be used unless the other dopes were unobtainable. *All machines after doping must be varnished on their upper surfaces with pigmented varnish P.C.10, and on their under surfaces with varnish V.114. This latter varnish (V.114) is on no account to be employed on the upper surfaces of any machines. If P.C.10 is not available, ripolin may be employed in lieu.*[7]

Both schemes may be seen in the accompanying photographs.

A Report of 19 August 1916, from Dover Seaplane Station noted that the P.C.10 *colouring material had peeled very badly* from the leading edge of a Sopwith Baby, and in doing so had removed all the dope from the fabric. This was also happening to a Short Single Seater Bomber Seaplane and *this colouring material is considered most unsatisfactory.* The Report noted that Messrs Short *have discontinued their practice of varnishing their planes, with the result that the fabric of the latest machines which have arrived at this station is already showing signs of slackness due to salt water and weather, although the machine has only flown for a few hours.*

The treatment of fabric by Messrs Short in the early part of the war was cited as being *highly satisfactory in every way.* Seaplane No. 1338 had done aver 200 hours without its fabric being renewed and apart from dirtiness and straining the fabric was almost as taut as the day it was delivered to HMS *Riviera.*

The Station had tried several experiments with the doping of planes and submitted the following procedure which *is the most satisfactory for the hard work that these machines have to contend with.*

1. A coat of half cellon and half acetone.
2. Rub down with dry doped fabric.
3. A coat of pure cellon.
4. Rub down with dry doped fabric.
5. A coat of pure cellon.
6. Rub down with dry doped fabric.
7. A coat of pure cellon.
8. Rub down with dry doped fabric.
9. A coat of pure cellon.
10. Rub down with dry doped fabric.
11. A coat of pure cellon.
12. Rub down with dry doped fabric
13. A coat of one-third varnish and two-thirds petrol.
14. A coat of one-third varnish and two-thirds petrol.

24 hours must elapse between each coat of dope and 36 hours before the first coat of varnish.

Special permission was asked for the machines attached to Dover and Dunkirk to be treated in the original manner as mentioned earlier in the report.[8] The Director of Air Service replied to this request and it was stated that it was not considered superior to that presently employed in the RNAS. It was decided to test some of the seaplanes at Dover in the manner described in the proposal and some in the ordinary manner and compare the results.[9]

Sopwith Baby No. 8195 was examined after 2 hours 15 minutes flying time at Dunkirk. *The fabric had been doped with the brown colouring matter material P.C.10, and is now literally hanging in bags on the upper surface of the wings.* The result of this was that the machine had to be sent back to Dover for refabricating after only a small amount of flying time, *instead of doing at least 60 flying hours before this is necessary for machines doped in the normal manner.* Dunkirk again asked that the machines sent be doped in the old manner. Dover had limited workspace and the experimental work should be carried at a station that was not an active service station. It was considered that *less time would be wasted if experiments were not carried out on machines which have to do continuous high performance work against the very efficient enemy machines which are being employed against us at the present time.*[10]

In September 1917, Short 827 No. 3097 was in the tropics and undergoing an engine change and overhaul.[11] The *engine cowling & undershields* were painted. Floats – main, wing and tail, struts and wires were also painted. This is evident in photographs. The colour is unknown but thought to be grey similar or the same as that applied to RN ships.

Sopwith factory drawings for the Baby seaplane refer to Ripolin Khaki. When Fairey rebuilt Sopwith Baby 8134, the upper surfaces were doped with Ripolin khaki A, and the originally clear varnished floats were painted light grey.[1] Sopwith Triplane fighters had, according to Sopwith drawings, P.C.12 covering dope applied to upper surfaces.[2] It is therefore thought that dark doped seaplanes may well have had this dope rather than P.C.10 applied, however, no documentation has been discovered to date to confirm or deny this.

National Markings

No. 890 was an Albatros B.II biplane purchased by the Admiralty before the outbreak of war. This aeroplane had an interesting life and spent some time at the Isle of Grain where it was involved in various experiments. On 17 December 1914, *In accordance with instructions in AD 5954 a red ring with a white centre is being painted on the planes in lieu of the Union Jack.*[12] This instruction superseded that on 26 October 1914, that had the Union flag marked on the under surface of the lower wings. It is assumed that this change came from the problems that troops had with identifying aircraft in that the flag looked like a cross when viewed from some angles. Most troops lacked any form of aircraft recognition and all aircraft were fired upon, but one with any marking that approached a cross came in for more 'hate' from the ground.

The RNAS used a red ring as a national marking with the Union flag often carried on the fuselage. The ring could have a thin white outline. It was not until 1 November 1915, that the RNAS adopted commonality with the RFC and adopted the tri-coloured cockade/roundel that was often referred to as 'circles.' Initially a blue centre was added to the red ring. Rudder stripes were adopted and sometimes were applied across the elevators as well.

End Notes:

1. Even Bruce Robertson's *WWI British Aeroplane Colours and Markings* (Albatros Productions, UK. 1996) barely touches on RNAS doping practices.
2. Notes from TNA AIR1/ 186/15/226/3 by J.M. Bruce. RAF Museum J.M. Bruce Collection Box 4.
3. Note for 28.08.1914. J.M. Bruce notes from TNA AIR1/186/15/226/3. RAFM J.M. Bruce Collection Box 4.
4. Log Book of No. 189. TNA AIR1/204/412/819.
5. Log Book of No. 832. TNA AIR1/204/412/842.
6. 'Temporary Memorandum No. 48.' *Memoranda* 1915. RNAS Air Department 1 January 1916.
7. Technical Memoranda 1916. Air Department No.19, issued 1 January 1917. TNA AIR1/2097/207/15/6.
8. CO Dunkirk to Senior Officer, Dover. 19 August 1916. Signed by Sqn Cdr F.E.T. Hewlett, CO. TNA AIR1/66/15/9/98.
9. C.R.J. Randell for Director of Air Services to CO Dover. 1 September 1916. AIR1/66/15/9/98.
10. Sqn Cdr F.E.T. Hewlett, to Senior Officer, RNAS, Dunkirk. 26.09.1916.
11. Log Book on Nos. 3097. TNA AIR1/2072/204/412/3097.

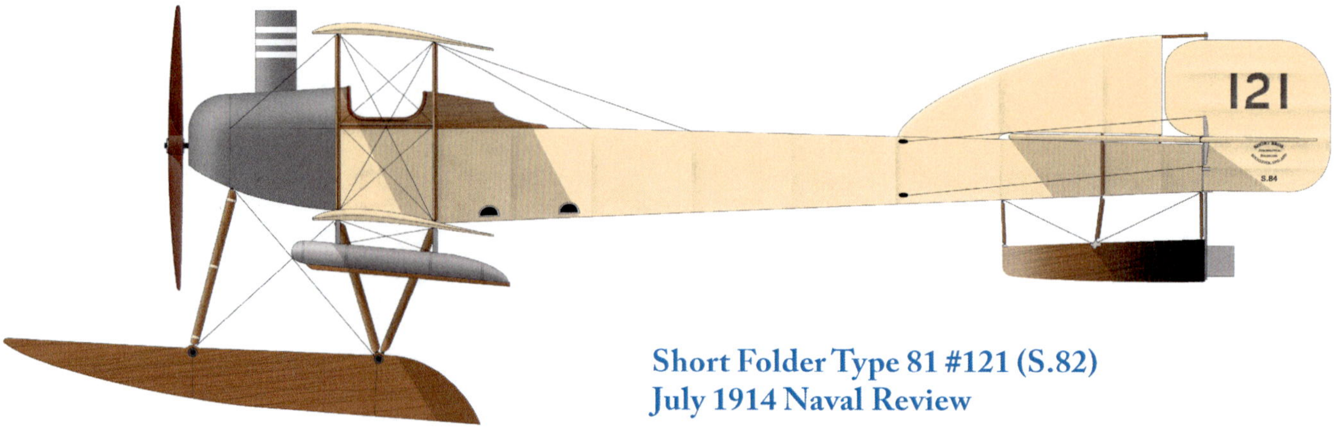

Short Folder Type 81 #121 (S.82)
July 1914 Naval Review

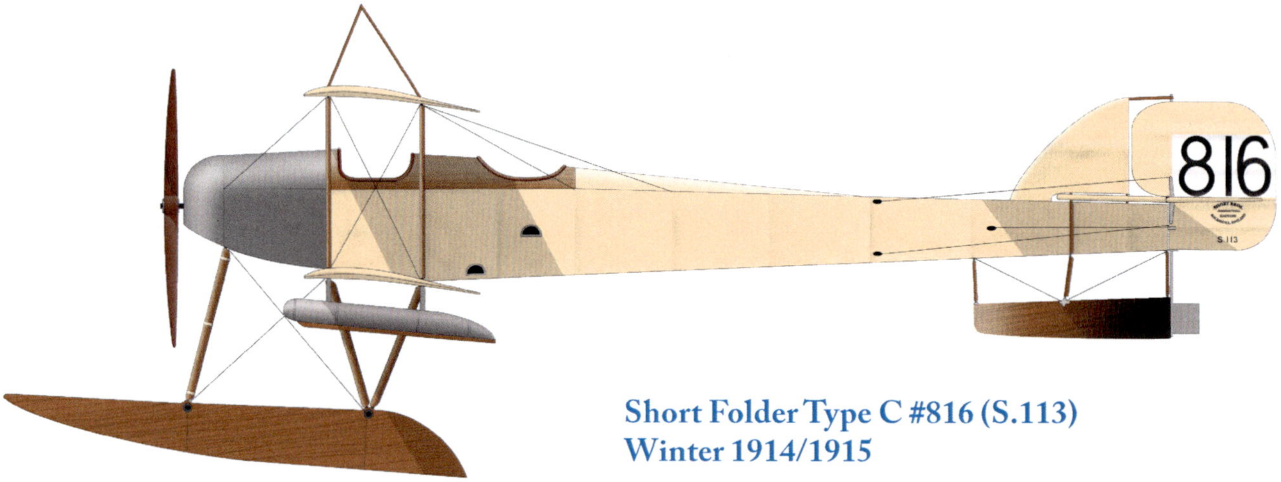

Short Folder Type C #816 (S.113)
Winter 1914/1915

Short Type 830 #1335

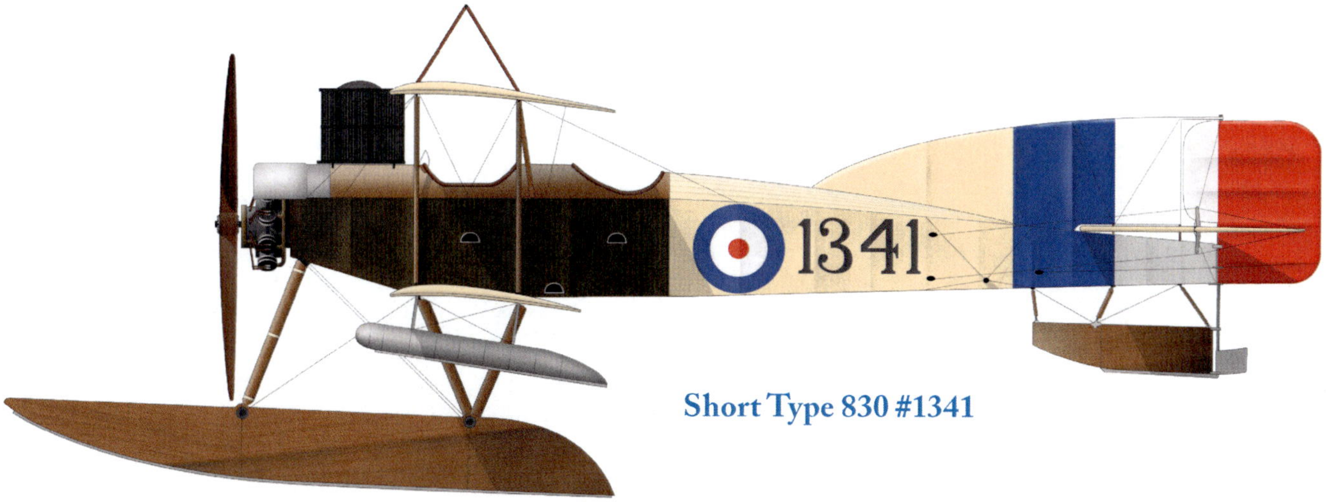

Short Type 830 #1341

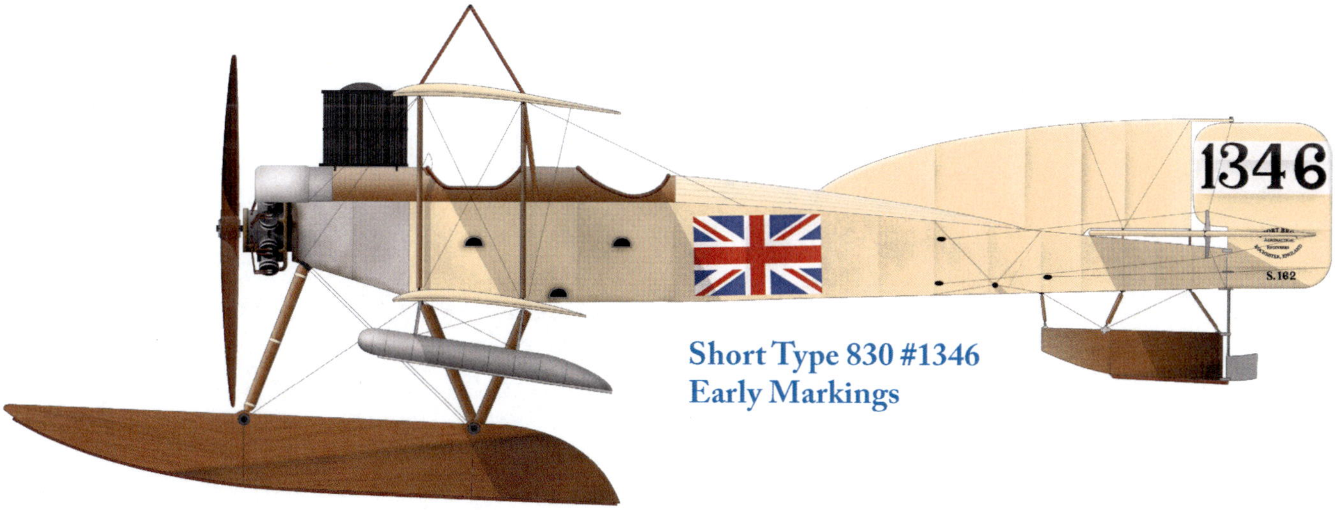

Short Type 830 #1346
Early Markings

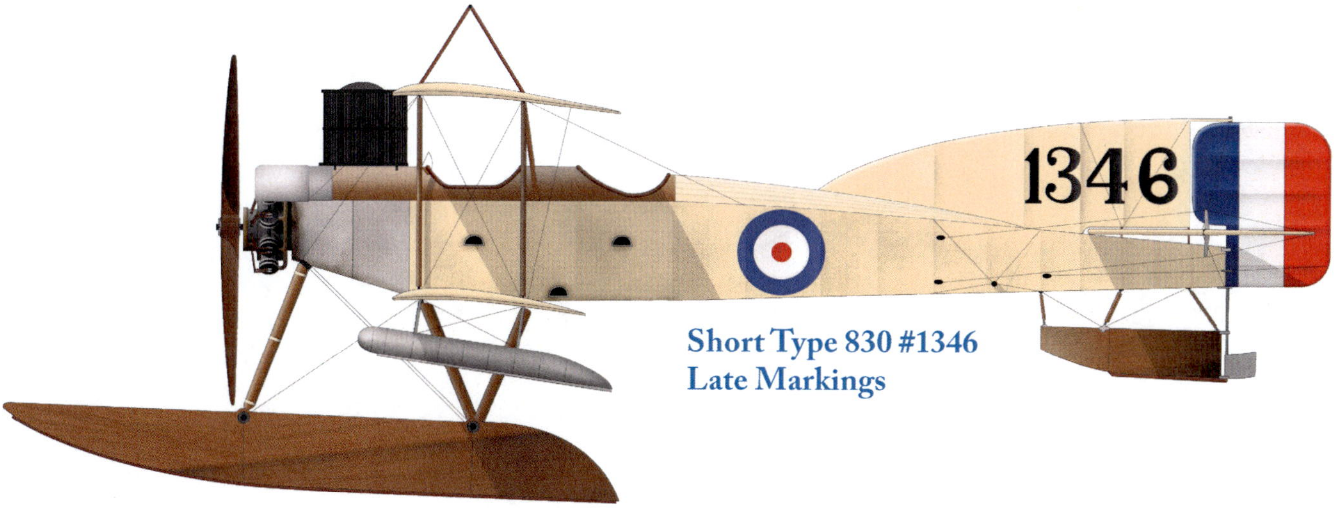

Short Type 830 #1346
Late Markings

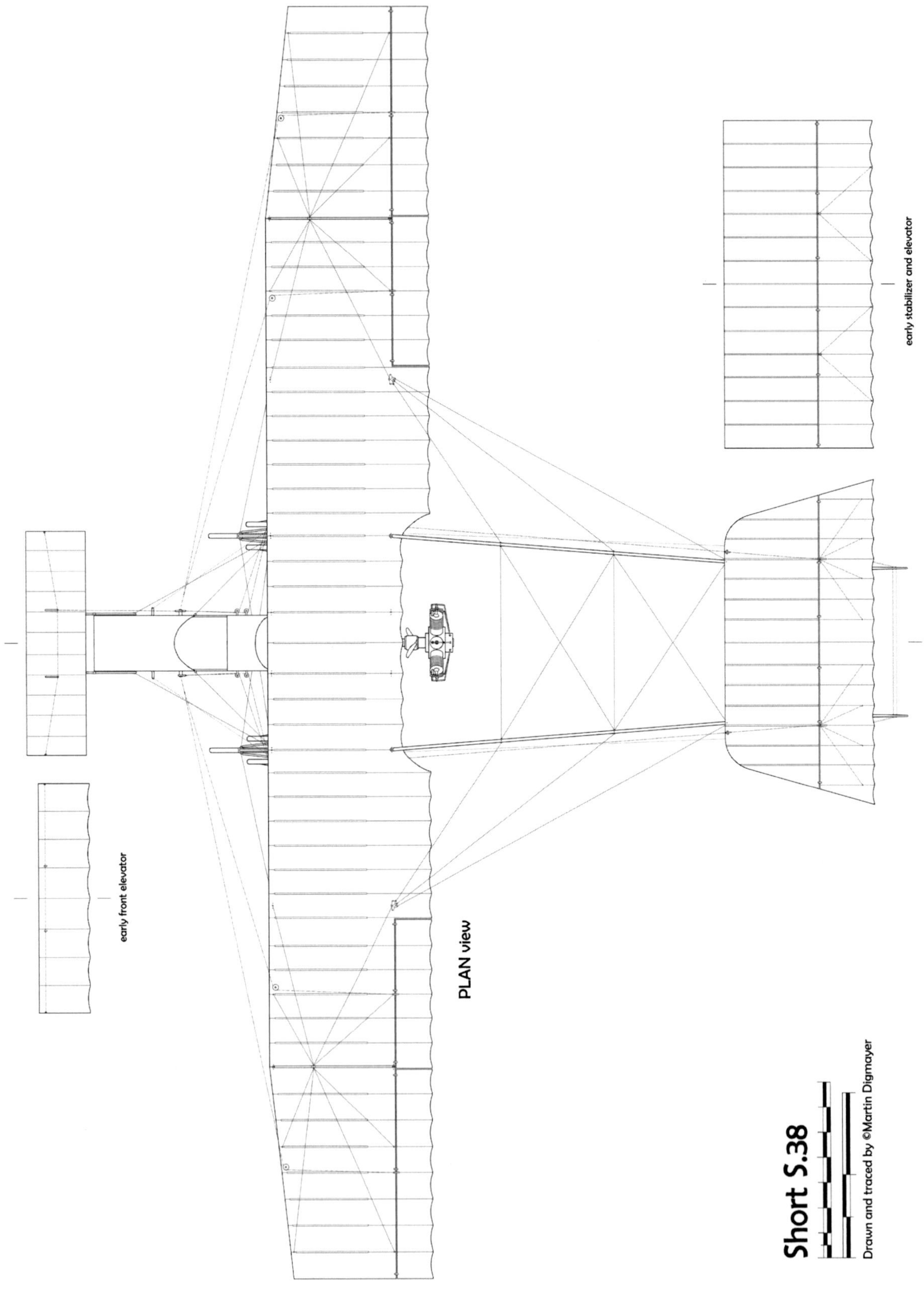
early stabilizer and elevator
early front elevator
PLAN view
Short S.38
Drawn and traced by ©Martin Digmayer

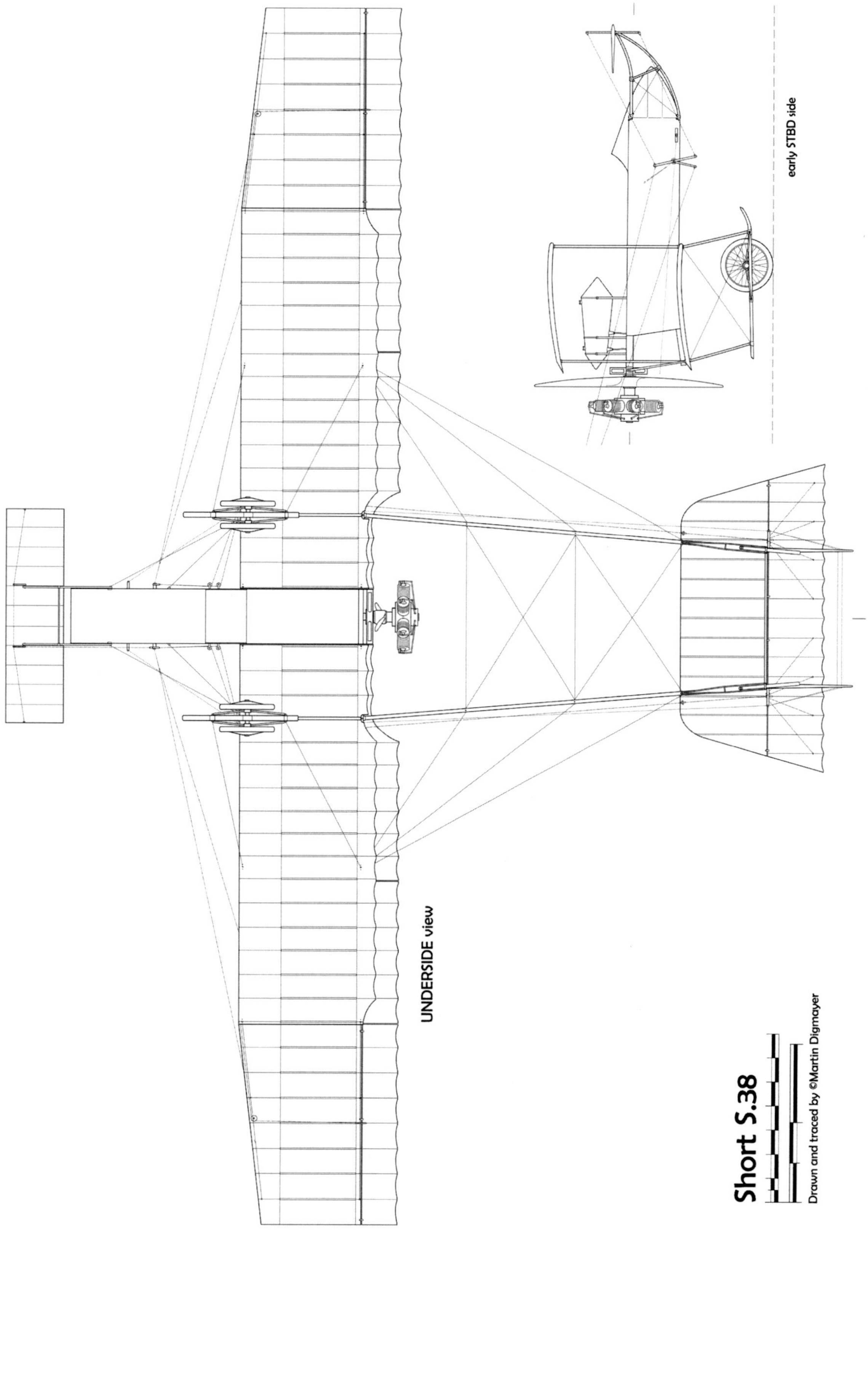
early STBD side
UNDERSIDE view
Short S.38
Drawn and traced by ©Martin Digmayer

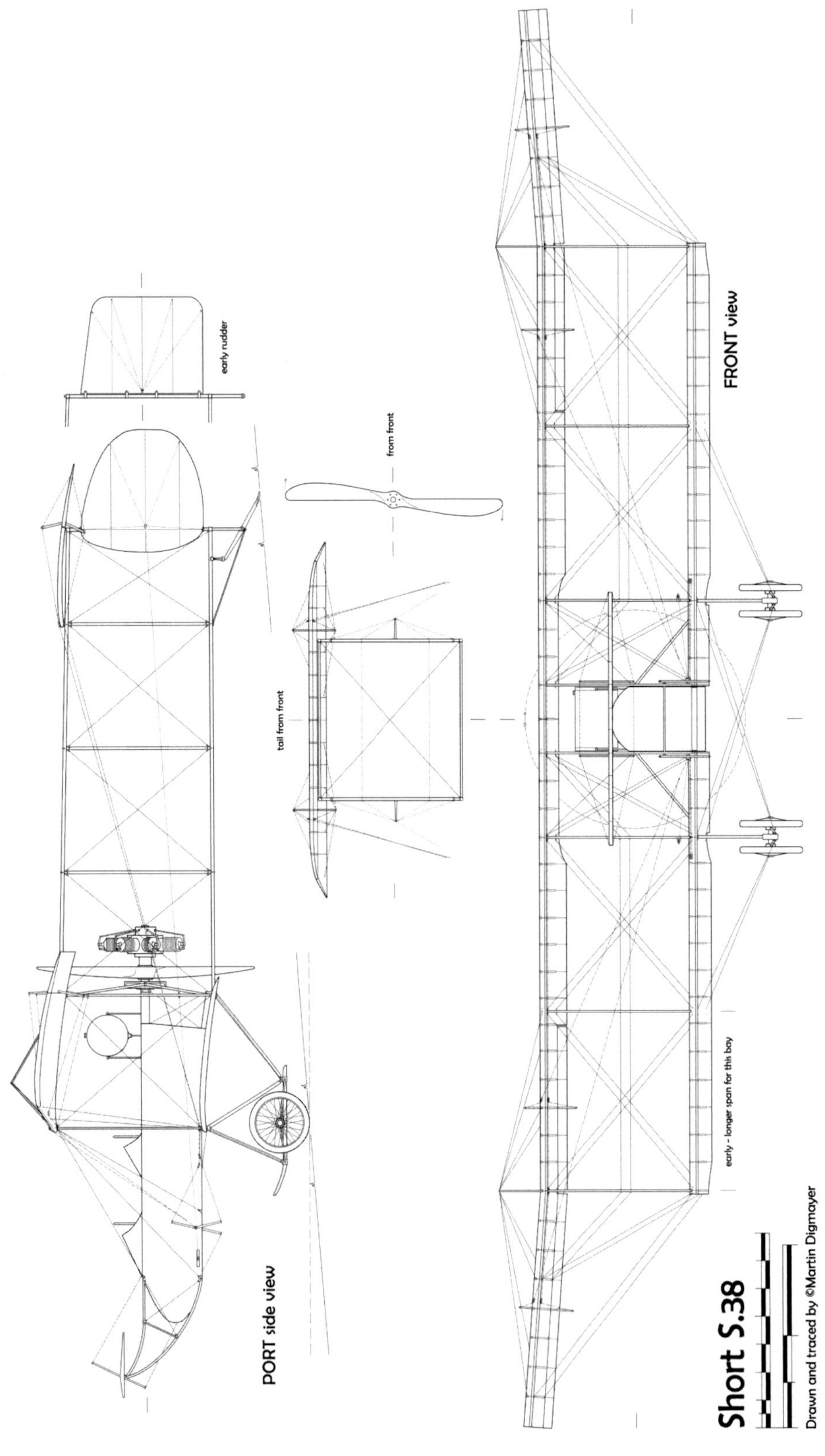
early rudder
from front
FRONT view
tail from front
early - longer span for this bay
PORT side view
Short S.38
Drawn and traced by ©Martin Digmayer

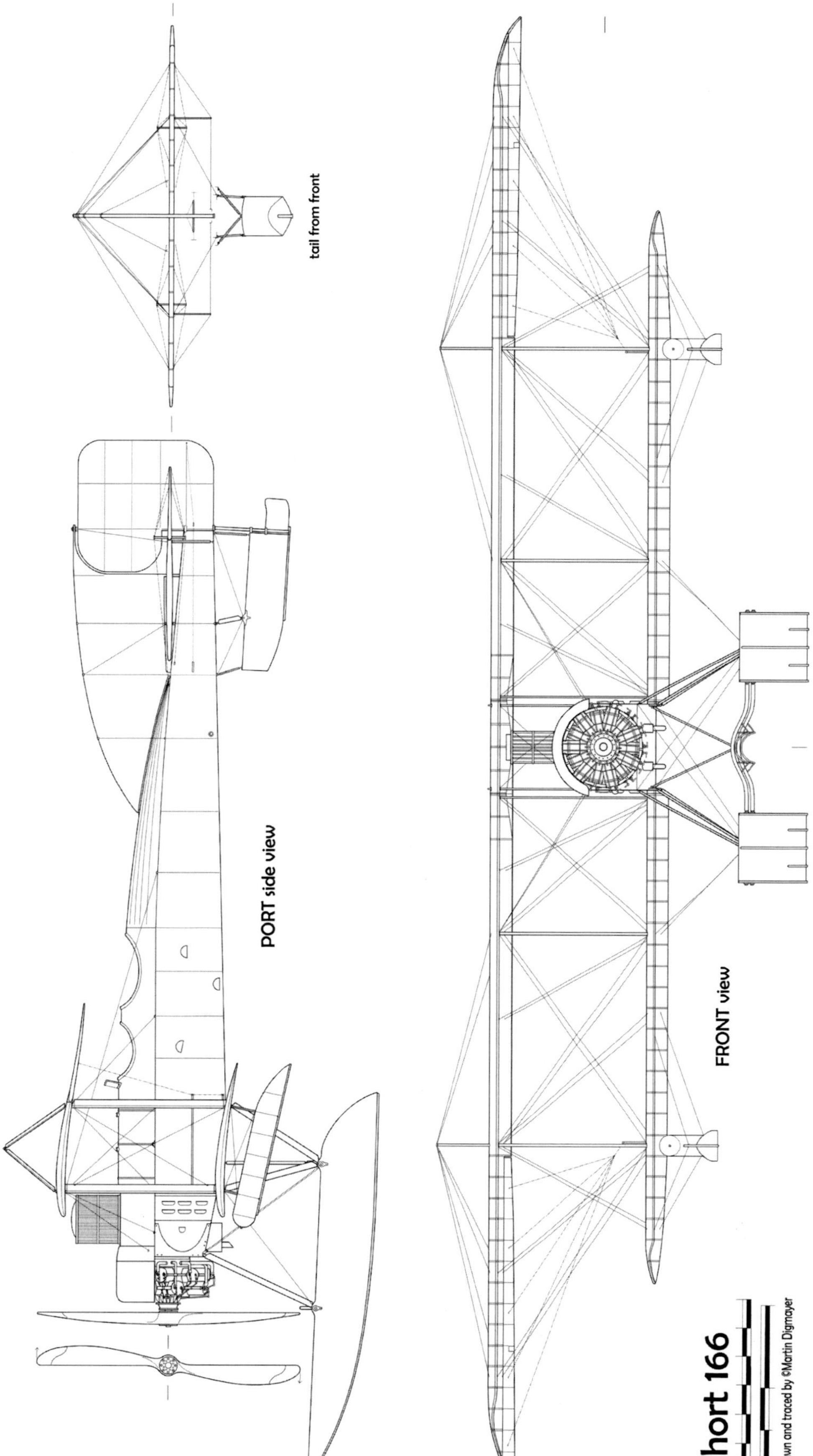
tail from front
PORT side view
FRONT view
Short 166
Drawn and traced by ©Martin Digmayer

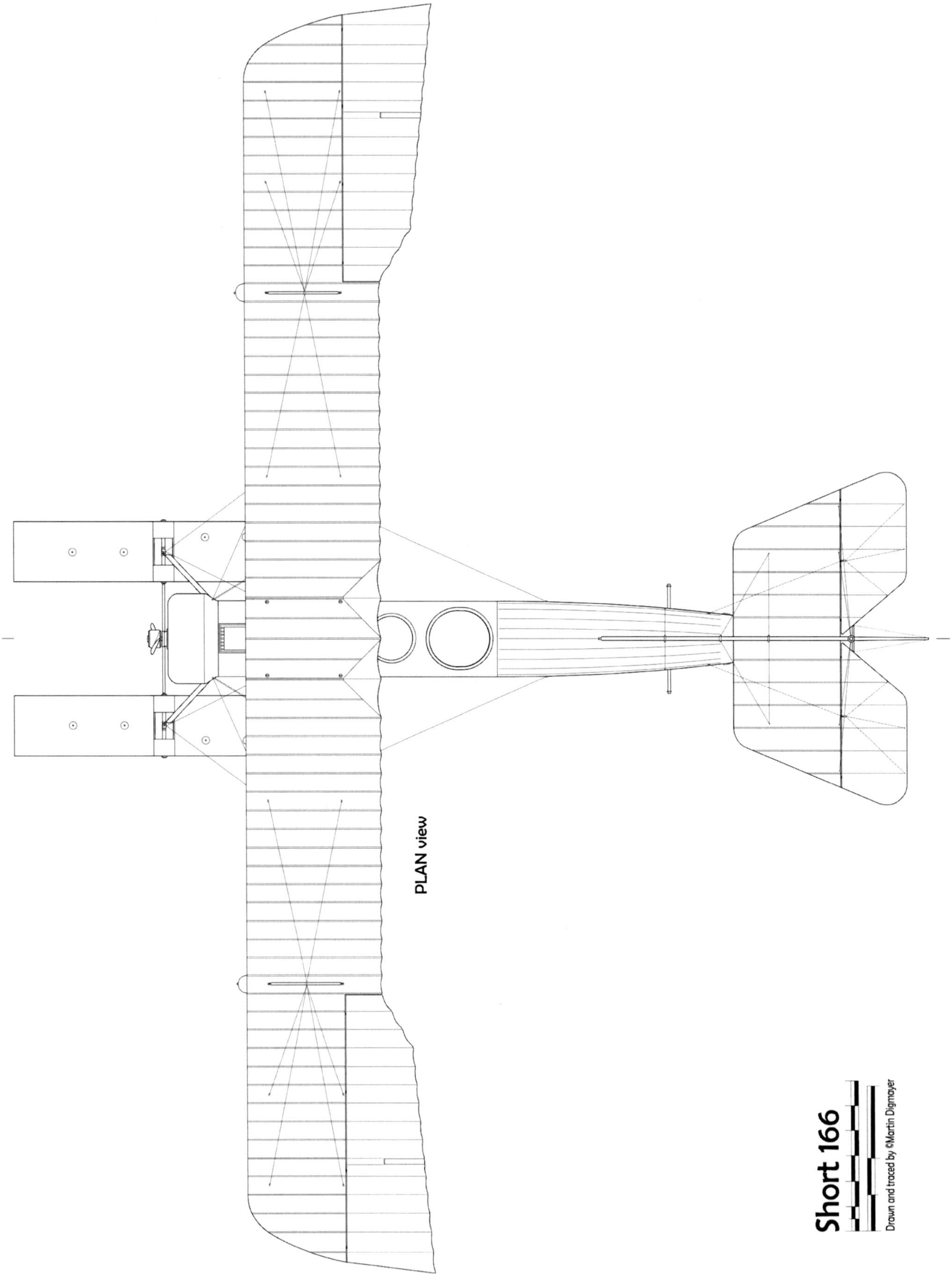
PLAN view
Short 166
Drawn and traced by ©Martin Digmayer

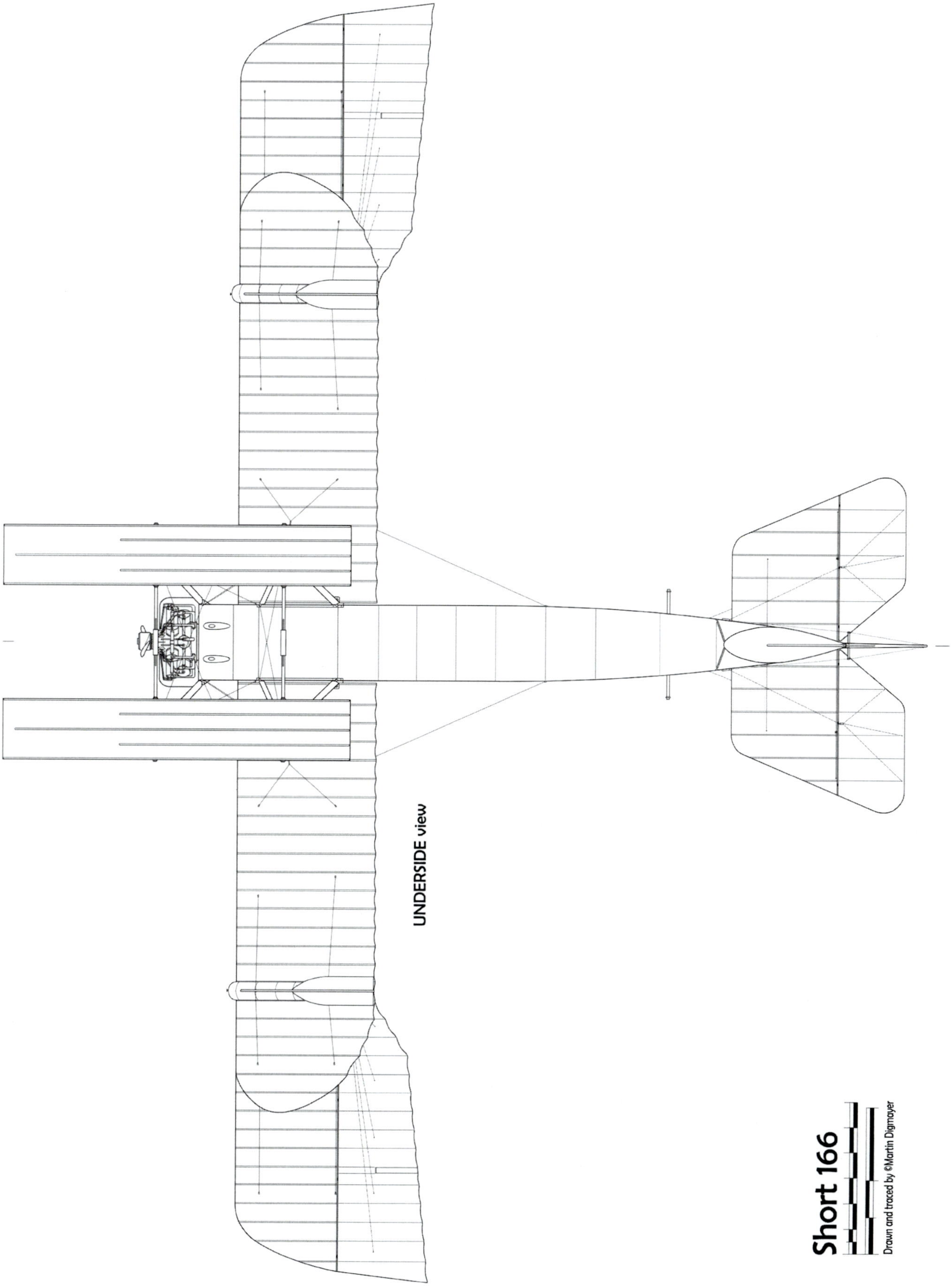
UNDERSIDE view
Short 166
Drawn and traced by ©Martin Digmayer

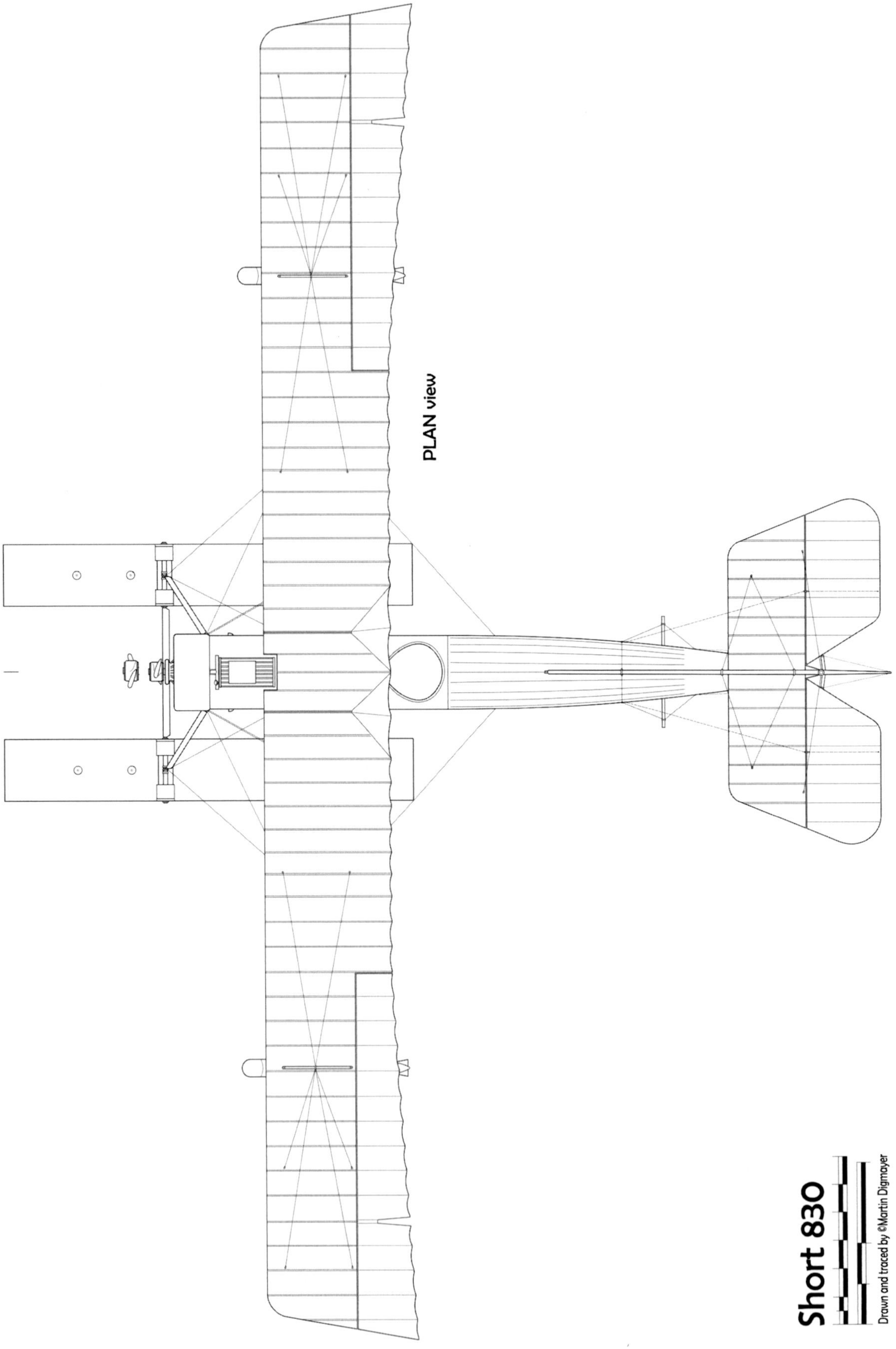
PLAN view
Short 830
Drawn and traced by ©Martin Digmayer

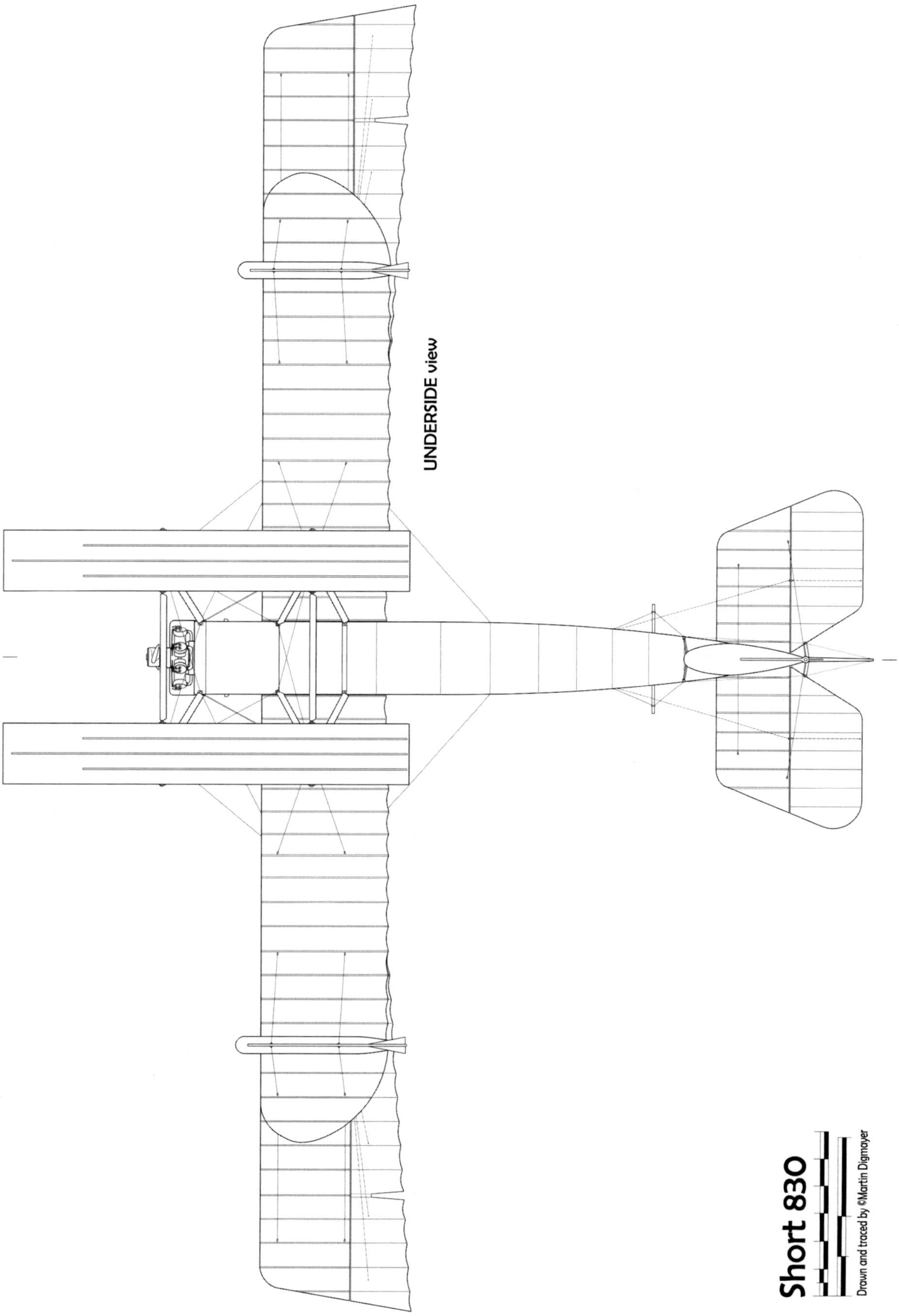
UNDERSIDE view
Short 830
Drawn and traced by ©Martin Digmayer

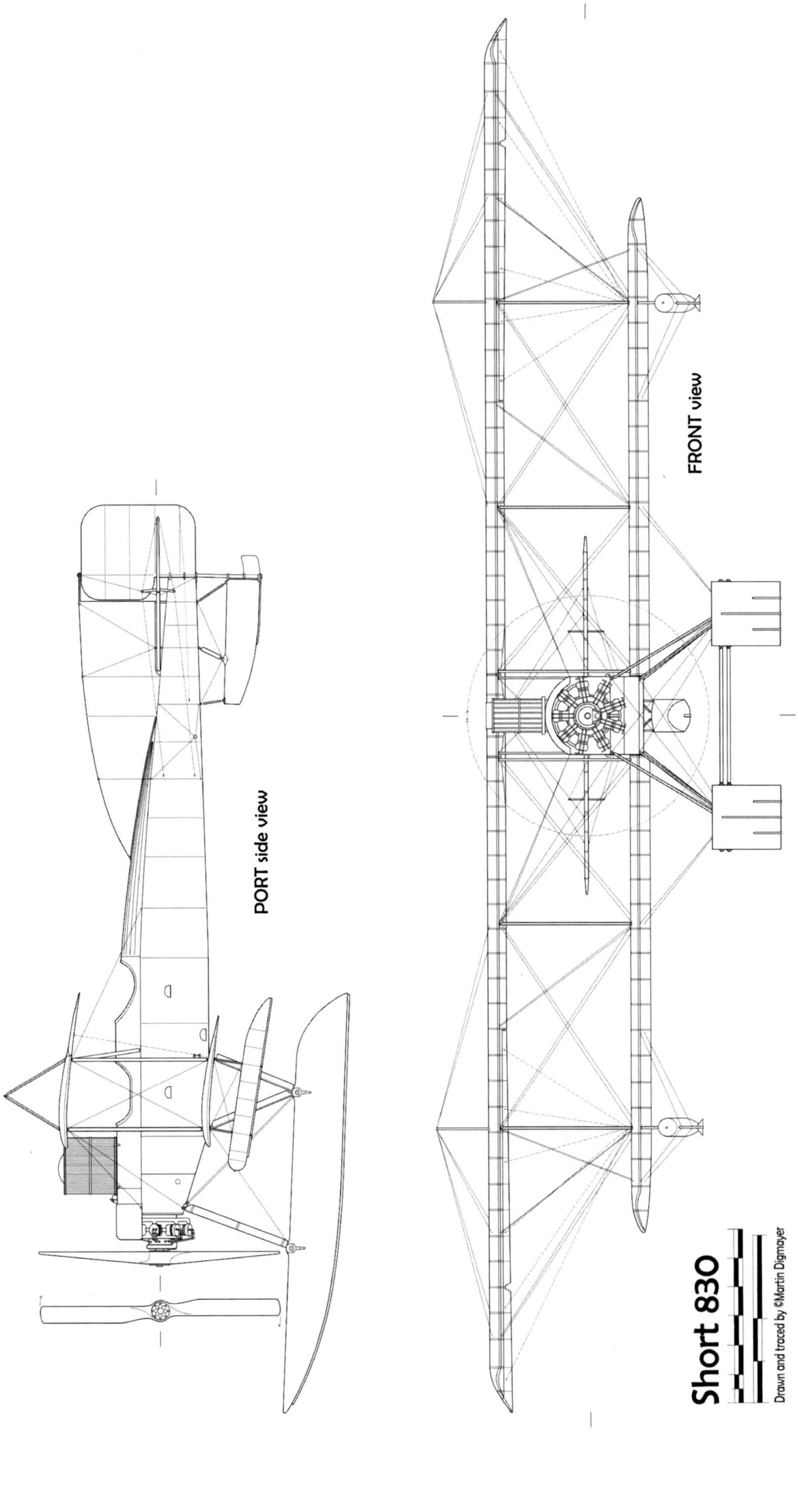
PORT side view
FRONT view
Short 830
Drawn and traced by ©Martin Digmayer

Made in United States
Orlando, FL
04 November 2026

90200384R00133